Why and How Toddlers Should Memorise the Times Tables

A guide for parents and teachers

Dr. Bijan Riazi-Farzad

www.mathtery.com/tt

Trexicon Publishing

London, United Kingdom

www.trexicon.com

Copyright © 2014 Dr. Bijan Riazi-Farzad

All rights reserved

ISBN-13: 978-1500994129

DEDICATION

To all my students and research subjects who have shared their hopes, fears, aspirations and frustrations with me and, in so doing, have helped me understand so much about the strengths and weaknesses of our education system.

and

To all my wonderful teachers who have inspired me to appreciate that the greater purpose of teaching and of learning is nothing more than for us to connect better.

A few quotations that I like

> "If I had asked people what they wanted, they would have said faster horses."
> — Henry Ford

> "It is not of the essence of mathematics to be occupied with the ideas of number and quantity."
> — George Boole

> "I am, somehow, less interested in the weight and convolutions of Einstein's brain than in the near certainty that people of equal talent have lived and died in cotton fields and sweatshops."
> — Stephen Jay Gould

About the Author

Eighteen years after he obtained his first degree in Applied Chemistry, and fourteen years after he started tutoring students on a one-to-one basis, Dr. Bijan Riazi-Farzad went back to university to study Psychology. He obtained his master's degree in Psychology of Education from the Institute of Education, University of London, in 2006 and became a Graduate member of the British Psychological Society in the same year. Since then, whilst continuing to tutor privately, he has been involved with research looking into the factors that affect children's attitudes towards their school subjects. For this work he has visited many schools, conducted hundreds of interviews with students and teachers and has been involved in the evaluation of several schemes designed to increase students' desire to study subjects, especially Physics and Mathematics.

Bijan completed his primary school in Iran before moving to the UK in 1975. He graduated in Chemistry in 1986. For his PhD, in pharmacology, he looked into how cotton workers develop an interesting type of asthma called byssinosis. He has been teaching science and mathematics in schools and as a private tutor since 1990. He has two primary-aged children and lives in London, United Kingdom.

Preface

I have a dream ...

My dream is for every child at school to develop a love of, and a lifelong relationship with, mathematics. Why? Because ...

Somewhere, beyond an invisible wall, there is a land of unimaginable beauty; a land where every citizen's purpose is to first appreciate the beauty inherent in every process and then to use that understanding to create more beauty; where it is everyone's resolve to share that understanding to improve the everyday experiences of every human being on the planet and to move humanity into a safer and more enjoyable future.

Those who have entered this land wish that everyone could join them. They want everyone to enjoy the fruits of the perspectives that citizenship of this land offers; perspectives that will allow everyone to appreciate more and to connect better.

I call this magical place 'Matherland', a place where Mother Nature and Mathematics meet. It is a paradise where fruits from the tree of knowledge are not forbidden.

Regrettably, most children grow up and die without ever experiencing this magical place.

Many people know of Matherland and lament that it is too late for them to take the journey; not because they cannot, but because they are now on a different journey. Another reason could be that they do not know the way to Matherland and have no guide to show them the way. Others may believe that the gates are only open to a chosen few. Unfortunately, such mindsets could deprive the next generation from the impetus, and therefore the opportunity, to take that journey. My hope is that this book will inspire you to help your children to start the journey to this magical land with strong foundations, including confidence.

Citizens of Matherland aren't just professors or scientists or doctors or engineers or actuaries or finance analysts, they are also more insightful plumbers, artists, philosophers, dancers, beauticians, politicians, teachers and parents. In fact, mathematical insights change the way that one perceives, and consequently enjoys and performs, in every endeavour.

In this book, I explain how certain perspectives within our education system and its consequent practices, including the way we in which we approach the teaching of the times tables, critically affect the quality of children's experiences of mathematics and how the approach we choose sets off a domino effect that could make or break a person's journey to Matherland.

My intention is to follow up this book with a series of other books to help you to give your children further support along the journey to that magical place of love and beauty.

About the Footnotes and Endnotes

One of the ways in which I establish rapport with my students is by thinking aloud as I consider how to explain something. It lets them see (or hear) how I am thinking the idea through, and what I am thinking about, as I try to find the best way of explaining it. If I just pause for a long time, my student may think that I am trying to think of what to say, but my 'thinking aloud' helps my students appreciate that I am thinking about, not *what* to say, but *how* to say it in ways that make the idea as easy for him[i] to appreciate as possible. So, by thinking aloud, I am able to convey to my student the sense that I am thinking hard to try to make the ideas as *relevant* to her as I can. The rapport that this creates makes it easy for me to sometimes say, "Trust me, I know why this is important for you, although I may not be able to explain why in a way that would be acceptable to you at the moment", without my student feeling patronized.

The footnotes and endnotes in this book serve a similar purpose. My intention is for you, my dear reader, to get a sense that I am trying hard to help you make sense of where I am coming from, so to speak. Some people may find it distracting and if you do, then I hope that I will have written in such a way that you will be able to make full sense of the main text

---

[i] Or her (and from now on, I shall use him/her and he/she interchangeably).

without referring to the footnotes or endnotes. Personally, I find that these notes enrich my experience of the content by adding extra context without interrupting the flow.

I have also used these footnotes as a way of connecting with you on a more personal level because listening to a friend is often easier than listening to a stranger.

There are some authors, like the contemporary Iranian author and historian, Prof. Bastani Parizi[i] whose writing is famous for having more footnotes than main content[ii] - and I don't mind that at all because, as I keep saying, meaning is derived from context: no context, no meaning; the more context, the richer the meaning.

I have assigned Roman numerals to the footnotes (shorter notes) and Arabic numerals to the endnotes (longer notes).

---

[i] RIP (April 2014)

[ii] One day someone, whom I know very well, called up Dr. Parizi and asked him if he could compile his footnotes into a book – apparently, he was very receptive to that idea.

TABLE OF CONTENTS

A WORD WITH PARENTS 1
   Understand the Principles 8
   Should Toddlers Learn their Times Tables? 10

INTRODUCTION 15
   The Child's Experience of Mathematics in the Classroom
   16
   The Seeds of Disaffection with Mathematics 19

SEVEN PRINCIPLES 23
   1 Memorising is a Prerequisite for Understanding 24
   2 Memorising the Times Tables is NOT about
   Understanding Multiplication 30
   3 Familiarity Brings Confidence 39
   4 Memorising and Understanding Require Different
   Learning Techniques 40
   5 Memorisation of Facts is Not the Same as Rote Learning
   42
   6 Memorisation is Not the Same as Recall 44
   7 Younger Children Are Better at Memorising Without the
   Need to Understand 46

SEVEN Rs (REINFORCERS) 51
   1. Respect 54
   2. Random 56
   3. Rapid Recall 56
   4. Rhythm and Melody 57
   5. Relate (Parallel Learning) 57
   6. Reward 58
   7. Repetition 58

THE WRONG WAYS OF TEACHING THE TIMES TABLES 59
'Pattern Seeking' Methods 62
Other Pattern-Based Methods.. 63
Story-Based Methods.. 64
The 'Look-up' method... 65
The 'Count-up' Method 66

THE TRIAD METHOD 67
Background 68
How the Triad Method   works 70

THE ANATOMY OF A MULTIPLICATION EXPRESSION 72

EXAMPLE EXERCISES 75
Introduction 76
Why a boat?... 78
Stage 1 Introduce the triad 81
Stage 2 Hide the Product 82
Stage 3 Hide the Multiplier 83
Stage 4 Hide the Multiplicand 84

THE 36 TRIADS 85

END NOTES 98

THE BLURB... 110

> *"Of all sad words of mouth or pen, the saddest are these: it might have been."*
>
> — John Greenleaf Whittier

> *"Don't let schooling interfere with your education."*
>
> — Mark Twain

> *"It is no measure of health to be well-adjusted to a profoundly sick society."*
>
> — Jiddu Krisnamurti

A Word with Parents[i]

---

[i] And teachers (being *in loco parentis*)

So, you have picked up a book about whether toddlers should learn their times tables. But why? Is it simply because mathematics is just another subject that your child has to study at school? I hope not.

Later, I explain how you can help your child to be much better at mathematics than she is likely to be without the right support. I say mathematics and not just the times tables, because trying to learn mathematics without knowing the times tables *off by heart*, is like trying to learn to play basketball with a medicine ball. That's why so many children find it difficult and ultimately give up.[i]

The exercises described this book *are* designed for your child(ren). However, to be effective, you need know what to do and how to do it. It would also be of great benefit to you if you know why the activities have been designed in the way that they have.

Don't worry if you are one of the majority of people who didn't fall in love with mathematics when you were at school. It's never too late. I hope that, as you progress through this series,[ii] helping your child to appreciate mathematics, you will also gain some insights into the reasons for your own negative

---

[i] Or at least give up trying to excel at the subject.

[ii] As I mentioned before, my intention is for this book to become the first in a series, covering other aspects of mathematics education (my next bone of contention is the way in which we approach the teaching of proportionality, which is often made far more complicated than it needs to be :-)

feelings towards mathematics and that this is no reflection on you, but on the 'system'.[1]

In this book, I advocate repetition as an effective means[i] for committing facts and ideas to memory. Unfortunately, drilling, or 'learning by rote', has been given a bad reputation in recent decades.[ii] As such, we are reluctant to teach our kids anything that we think they won't understand, lest we be thought of as renegade.

Children have no problems with repetition. As parents will testify, young children can watch the same television or video programme over and over again without getting bored. The reason is that they are looking at it with a fresh eye every time.

Often, with 'drilling', what they complain about is neither *what* we are trying to teach them, or *why*, or even *how*; but the context. Let me repeat that, because it is crucial; especially for younger children, **the context is more important than the content, the purpose or even the method**.[2]

Many educationalists consider mere memorisation, as fundamentally flawed because they believe that children should not be 'taught' what they cannot

---

[i] Although, as I explain later, not the *only* means

[ii] And isn't it curious that Carol Vorderman, David Cameron's new 'maths tsar' says that at school, "The times tables were drilled into us." My argument here is that she wouldn't have the reputation she has now if they hadn't been.

understand.[i] I believe that this view is, at least partially, responsible for what has been referred to as the dumbing down of education in recent decades.[3] I consider 'mere' memorisation to be an effective way of building the foundations on which higher level understanding can develop. Let me explain.

Much of this resistance to memorisation prior to understanding could stem from the misapplication of Jean Piaget's theory about the various stages of cognitive development in children.[ii] The point that we are missing is that Piaget's work does not imply that children should not memorise what they do not understand; only that, before a certain stage, they will not understand what they memorise. Even then, it is not that they will not understand it *at all*, it's just that the way they make sense of it is *qualitatively different* from the way they would do at a different stage. I will say a little about more empowering interpretations of Piaget's stages of cognitive development a little later on.

Children have an enormous potential[4] to absorb and memorise. If we assume that we should not teach them about ideas until they are ready to understand them, we leave their vast potential to learn at the mercy of, at best, random events.[5]

---

[i] In the way we, as educators, expect them to understand.

[ii] Essential reading for trainee teachers.

Those who excel at mathematics have better prospects. Of course, you could argue that those who excel at *anything* have better prospects; so why single out mathematics? Because mathematics is about learning new ways of thinking that challenge and change the way we look at every subject. Yes, you could argue that history, geography, philosophy, psychology, dancing and even watching TV, can all do that to a lesser or greater extent, but mathematics is different.

In our research[i], we asked tens of thousands of twelve and fourteen year old students what they thought mathematics was about and, tragically, the most common answer was, "Mathematics is about *numbers.*" **Please don't let your child end up thinking that way!**

Yes, the focus of this book is on the times tables, but, through my explanation of why memorising the times tables is important, I hope to impress upon you that mathematics is as much about numbers as language is about letters of the alphabet.

Mathematics is about how any one thing is related to any other. That is to say, it's about *relationships.* In fact, **mathematics is THE language of relationships.**[6] It's complementary to, and not a

---

[i] At the Institute of Education, University of London

substitute for, all languages. This is why it is essential to the study of almost everything, even philosophy.[7]

Knowledge is not an end, it is a means.[i] The foundations that we lay down for *each* of our children will affect the prospects and the futures of *all* of our children and *how* we teach should depend on *why* we teach and not necessarily on what we teach.[ii]

I, therefore, beseech you to seek to know and understand *what* our children are expected to acquire[iii], *why,* and just as importantly, *how.*[iv] Yours and all our children's futures depend on it[v]. Don't allow influences outside of you to make you feel helpless to make a difference. You may not realize it, but every one of us is making an enormous difference every single moment,[8] some by taking action and others by being passive.[9]

Yes, I know you are busy and you haven't got the time and that's why you delegate your child's education to the professionals. And if that is the reason, then fine. On the other hand, if you knew that by finding out a

---

[i] It doesn't matter how much a child – or an adult for that matter – knows; what matters is how it will help them to grow, contribute and ultimately, feel fulfilled.

[ii] In the words of Steven Covey, in his best-selling book, Seven Habits of Highly Effective People, "Begin with the end in mind".

[iii] by different vested interests

[iv] Too often passively and hence, lacking self-direction.

[v] Since the future of each child is interconnected with the future of every other child (holism, interconnectedness, the butterfly effect, etc. (and/or read "I, Pencil" by Leonard E. Read).

little about educational principles and the extent to which they are being applied to your child's education, you could make a significant difference to your child's future, could you find the time? That's all I am asking.

Some of you may be wondering why, in a world where computers and calculators are everywhere, anyone should care about the times tables. And this is a very valid point. Mathematics teachers instinctively know why. That is to say, they have a gut feeling about the relationship between the ability to recall the times tables rapidly and the mathematical performance of the youngsters in their trust. However, it is very difficult to put such instincts into words in such a way that can convince parents of the value of making the memorisation of the times tables by their child a priority. Often, the argument that teachers make for parents and their pupils is that memorising the times tables is *better* (faster) than using calculators. This is not convincing enough for the harried or sceptical parent to focus relentlessly on making sure that their beloved child can recall her times tables as easily as the names of her favourite cartoon characters.

Here, I am attempting to make this argument more convincing so that you can enthusiastically, and confidently, put the memorisation of the times tables at the top of your agenda for your little ones.[i]

---

[i] And even the not so little ones (better late than never)

Understand the Principles

There are a lot of books that are written to help children with their times tables. These often consist of going through times tables grids in sequence. Many also try to *explain* or *show* how the times tables work by highlighting the patterns, especially the patterns of repeated addition. My contention is that, in spite of the popularity of such methods, **teaching the times tables in this way is counterproductive**.[10]

I feel that it is very important for parents[i] to become **actively involved** in understanding educational principles. Far too often, parents are given material to simply hand over to their children without any guidance as to how to make sure that the material is used **in the most effective way**. As parents, we need to know not just *what* to teach, but *how best* to teach and, just as importantly, *why*.

You might then ask, "But isn't that what the teachers at school are supposed to do?" And my answer would be, "If you wanted to leave it all to the school, you wouldn't be looking at a book like this." Secondly, think of home and school as members of the same team aiming to reach the same goals. The more you help your child's teachers, the more those teachers will be able to[ii] help your child. And, *if* you want to

---

[i] Yes, that includes you 'too busy' mums and dads out there
[ii] And be motivated to

give your child some extra help at home, then you need to know how to decide which resources are best[11]. I have written these notes for parents for several reasons, including,

1. To impress upon you the importance of memorising the times tables at the earliest possible age.
2. To persuade you not to leave the task of teaching the times tables entirely to your child's school.[i]
3. To encourage you to think of working through these exercises (or any other academic endeavour) as an excuse[ii] to have lots of fun with your child(ren).
4. To reassure you that the ideas presented in this book have been carefully thought through and are based on sound research, as well as many years of personal experience.
5. To broaden your perspective to consider the principles that you learn here as a means of personal development, not just as a parent, but in other areas of your life too.[iii]

---

[i] Remember, work with your school as a team.

[ii] If the word 'excuse' doesn't sit comfortably with you, replace it with. 'opportunity' or 'platform'.

[iii] For example, wouldn't be useful to know how your memory works? Or how your possibly poor relationship with mathematics may have initially developed?

The overarching purpose for all these; I call them *the whys*,[i] is to **motivate** you. Motivation is that sense of purpose that gives us that extra push to find the time (even just five minutes) in our busy schedule to sit down with our children to direct their attention towards something that we believe is better for them, where otherwise, without that sense of urgency and purpose, we might just leave our beloved to their own devices because we are simply 'too tired'.

Should Toddlers Learn their Times Tables?

There is one central message in this book; Yes! Children should learn their times tables as early as possible. The rest of this book presents arguments to convince parents, teachers and policy-makers of the importance of this.

If you say to a toddler, "Incy Wincy, ..." she will happily, and often proudly, finish that phrase with, "spider". Incy and wincy have no meaning. The toddler has probably never seen a spider and, to her, the pictures of spiders in books are just interesting patterns, just as the shape of the number 8 is an interesting pattern. It makes no difference to the toddler whether the triad is "incy wincy spider", "humpty dumpty sat on a wall", "twinkle twinkle little star" or "seven eight fifty-six".

---

[i] And consequently, *wise* :-)

Now, fast forward a few years. The child is now at primary school and is asked, "what's seven times eight?" and she begins to say, "errr, uhm, oh, let me see... two times seven is, erm, 14, and twice that is twenty eight and twice that is, erm, let's see, let's partition the 28 into 25 and 3, then two times twenty five is fifty and two times three is six and so, yeah..."[i] and with a sparkle in her eyes and proud of her achievement she says, "It's 56 miss." Of course, she is rewarded for understanding the principles and being able to 'think through' the problem. And I *do* believe in rewarding the child for doing this, because it is the way that she has been encouraged, or taught, to approach 'the problem'.

However, I also think that this is a tragedy as far mathematics education is concerned, because the times tables should not be treated as 'a problem' to be solved in order to arrive at a 'solution'. The times tables should be treated as FACTS. They are tools that are used for solving other problems, they should not be problems in themselves. Let me reiterate, **the times tables is**[ii] **a tool**; in the same way that a knife is a tool. If you need to use a knife, you would not be expected to know how to make one, or even how it is made, in order to use it. That's what tools are for; to

---

[i] Some of my friends find it hard to believe that this is really what is going on in our kids' heads and in our schools, but ask the teachers, they'll tell you.

[ii] Yes, I know, I am having problems deciding whether the "times tables" is singular or plural, so I'm going to treat it as a plural noun (like family, where the verb form used depends on the meaning that is being conveyed).

be used not to be made; or in the case of the times tables, "understood" or "derived". That will happen in time, *if the situation demands it*. But right now, I need to be able to peal this potato and I need the knife. I don't need to know where it came from or how much it cost or how it was made. We need to treat the times tables like that and I have written this introduction to the exercises and the sections that follow with a view to convincing you of that.

Here's another analogy. Would you expect a child to work through a 'problem' to work out that the things that stick out of trunks of trees are called branches? Imagine asking a child, "What's the thing that that bird is sitting on?" and she says, "Oh, oh, I know... it's the thing that <u>br</u>eaks, so it starts with 'br' and it rhymes with 'ranch' because animals eat the leaves off it ... Oh, I know, I know, it's a 'branch', yeah".

As ludicrous as that sounds, that's what we do when we expect children to 'understand' the concept of multiplication (or even number) before they learn their times tables. Just because somebody once understood number before they devised the times tables, it does not mean that the child memorising the times tables needs to understand that process.

"So what?", I hear most of you still wondering. "Nobody *needs* to know their times tables these days anyway. Calculators do everything for you." And that's where I'd argue that we've got it all wrong.

In the days before calculators, the times tables served two purposes; as a means of doing mental calculations and for second purpose. It is this second purpose which is being overlooked.

We do not learn the times tables to be able to multiply; not any more. I agree with you. That's why I have written this book; to impress upon anyone who is affected by mathematics education[i] that the times tables should be treated as a tool that will make it much easier for children to understand what mathematics is *really* about, WHEN THE TIME COMES.

Mathematics is about exploring new ideas and new ways of thinking about relationships.[12] Without rapid mental access to the times tables, the initial learning curve becomes much steeper and it puts students off. This denies students from the joys and benefits of mathematical thinking throughout their lives.

Let me explain …

---

[i] And let's face it, that's all of us, in some form.

> "Every good mathematician is at least half a philosopher, and every good philosopher is at least half a mathematician."
>
> — Friedrich Ludwig Gottlob Frege

After learning the First Proposition, a youth who had begun to read geometry with Euclid (of Alexandria), asked, "What do I get by learning these things?" So Euclid called a slave and said "Give him threepence, since he must make a gain out of what he learns."

> "A ship in harbour is safe, but that is not what ships are built for."
>
> — John A. Shedd

Introduction

The Child's Experience of Mathematics in the Classroom

If you are asked to find out how much you have to pay for 8 chocolate bars costing 45p each, there's no reason nowadays to do that in your head. All you have to do is to take out your calculator and bingo, the answer is £3.60; done.[13] The problem starts a little later, when we are trying to teach the principles of more complicated mathematics. It is then that we need to use examples that require your child to be able to do simple multiplication quickly. This is because she has now reached a stage where she can, and needs to, focus her attention on making sense of the underlying principles. If she has to take out her calculator or to look up tables several times just to attempt a simple problem, the purpose of which is to teach her something that has *very little to do with the actual numbers*, that's when the slow pace becomes a **big problem**. It leads our beloved children to lose track of the main idea and become frustrated because they do not understand what is going on, and worst of all, become *needlessly* disillusioned with their mathematical ability.

The difference is in the perceived purpose of mathematics and is roughly the difference between primary and high school mathematics.[14]

Let me make this a bit clearer.

After children learn about the idea of number and the four basic operations[i], they can advance to the next stage

---

[i] Addition, subtraction, multiplication and division

where they begin to *use* mathematics to *derive* something that is not known from some information or relationship that is already known.[15] For example: ☐ x 5 = 35. For this, the child can look up the number 35 in a times table grid and find the answer; which is 7. Even at this stage, the faster the child can do this, the faster she progresses to the next stage and the more confident she will feel.

Now, an important milestone is the stage *after* that. To illustrate this, I am going to compare two students, Sam, who knows her times tables and Alex, who does not. Let's see what happens.

The teacher presents the children with this problem:

"If $5x = 35$, what is x?"

The teacher explains that this is similar to saying, if five somethings (let's say apples) costs 35 (let's say, pence) then how much does one apple cost? She tells the children that the answer is 7p. They are then asked to try out similar questions/problems.

In the following table, the first column shows the problems that the children are being asked to solve. In the second column, we see what goes on in the mind of Sam, a student who has memorised the times tables. In the third column, we see what Alex, who has not memorised the times tables, is thinking.

The Problem	The thinking process of Sam (knows times tables)	Alex's thinking process (does not know times tables)
If $5x = 35$, what is x Teacher's answers: 7	"Oh! Five 7s are 35, I know that from the times tables. So the x is what you multiply by 5 to get 35. I can do this." And moves on to the next question.	"How did she get that?" After some time and may be some hints from teacher or peers "Oh! I have to look it up in the times tables. I see." By this time Sam is several questions ahead.
If $8x = 48$, what is x	"I know that. Eight 6s are 48, so the answer is 6. I can do this." And moves on to the next question.	"Now, where is 48 on this grid. OK, there's the eight and yep, the other number is 6. So, what I have to do to answer these questions is to keep looking them up on this grid."
Ten questions later, they get to the last one …		
If $14x = 182$, what is x	"For the questions I did before, I knew what times the number gave the answer, but 182 isn't in the times tables so… How do I do this? What would I have to do with 35 and 5 for that other question to get the answer 7? Oh, yeah, I can divide 35 by 5 to get 7. Hey, I get it. Something times 14 is 182, that means that I have to divide 182 by 14 to get the answer." "Miss, Miss! I've finished and I get it. If something times x is the answer, then the answer divided by that something gives x. I'm a genius!"	Alex is still on the third question. But even if he does reach this question, he is stumped because 182 is not in the times tables grid.

When Alex hears Sam say, "I've finished and I get it," he thinks, *"What? How did she get that? And how did she do all those questions so quickly?"* But unfortunately, **Alex's reaction does not end there.** Alex now says to himself, *"I must be no good at maths. I must be dumb"* or may be more defensively,[16] *"I'm not a geek like Sam."*

The Seeds of Disaffection with Mathematics

"A Strange Inversion of Reasoning" [i]

Now, since Alex has noticed that he is not as fast as some of the other kids, he decides that he is slow *because* he doesn't get it. But in reality, it's the other way round; **he doesn't get it *because* he is slow**. He is spending so much time, and his mental energy is so focused on, the mechanical process of *looking up*[ii] the answer, that his mind doesn't have the time to step back to see the bigger picture.

Crucially, this is a bigger blow to Alex's[iii] future prospects than it might appear from just one incident because Alex has now started on a journey down that slippery and self-fulfilling path of diminishing self-confidence in mathematics. This is because, once

---

[i] Quote inspired by a talk by Daniel Dennett at TED

[ii] Or counting up – see later

[iii] And for society as a whole

this seed is planted, another[i] well-known psychological phenomenon called *confirmation bias* [17] kicks in. From then on, Alex will notice (his subconscious mind will *selectively* bring to his conscious attention) all the 'evidence' in favour of (confirming) his *inability* to do maths and he will ignore all the evidence that suggests that he *can* do it.

The point I am making is that the slower students appear to struggle (be slow), not because they *can't* do it, but because they are using *the* **wrong tool**, that is, looking up or counting up, rather than retrieving the answer directly (and quickly) from their memory. Not being aware of this, they falsely think that being *slow to respond* means that they don't have the *ability to understand*.[18]

One of my students, 15 at the time and working towards his GCSEs, told me, "I am not fond of maths" and was uncomfortable around anything that looked like maths. For example, whilst he liked science, he avoided those questions where manipulations of number was involved.[ii] Unfortunately, this put much of Chemistry and most of Physics, out of his reach. I, therefore, decided to put science aside and focus on mathematics. I found that he was struggling with mental maths and so I went back to basics, beginning with two and three digit multiplication and division.

---

[i] The first one was the psychological defence mechanism (see End Note 16)

[ii] Such as working out the yield in chemistry

What I noticed was that he was pausing a great deal between writing down digits and that was because he was using the count-up method to multiply.[i] This meant that, for example, if he had to multiply four by six, he was counting in fours six times, "four, eight, twelve, sixteen, twenty, twenty four", before he was able to go to the next step. No wonder he found it tedious. It was also no wonder that he did not feel confident with mathematics. He had most probably experienced the same feelings as Alex.[ii] My homework for this pupil was to go through the mechanical process of multiplying two and three digit numbers until he stopped counting up. After that, as he stopped being self-conscious about being slow at simple operations and began to pay greater attention to the underlying principles in topics such as percentages, ratios and algebra, his confidence grew.

To put these ideas into perspective, in the next section, I am going to discuss what I have called 'Seven Principles' that need to be borne in mind when *devising* methods for teaching the times tables.

In the following section, I will discuss what we need to consider when we want to put what we have devised on the basis of these principles into actual practice.

---

[i] More about this later

[ii] Although the reasons would have been long forgotten and only the feelings will have remained, which is even more frustrating because he wouldn't know why he does not feel confident with maths and attributes it to his 'ability'.

> "Do not worry about your difficulties in Mathematics. I can assure you mine are still greater."
>
> - Albert Einstein

> "The study of mathematics, like the Nile, begins in minuteness but ends in magnificence."
>
> - Charles Caleb Colton

> "If I were again beginning my studies, I would follow the advice of Plato and start with mathematics."
>
> - Galileo Galilei

Seven Principles

1
Memorising is a Prerequisite for Understanding

I am going to debate this at length because this is important and it is often misunderstood; regrettably, to the detriment of our children.

Children always ask, "What's that?"[i] *before* they ask, "What's it for?" Younger children don't even ask "What's it for?", but they still ask, "What's that?". They instinctively know that they need the labels before they begin to form the relationships between those labels.

Think of the times tables as 'labels' not as 'concepts'. When they can repeat, yes, parrot-fashion if you like, "seven eight fifty six", there is no reason at this stage for them to understand what that means. In fact, I would go as far as to say that it is unhelpful to try to get them to *understand* what it means at the same time as when we want them to memorise it; it slows down *recall*.[ii]

Coming back to our original question, why should we put our precious kids through the torture of memorising their times tables?

---

[i] That is, give me a (yes, an initially arbitrary) label to attach to this object.

[ii] And it can be counterproductive too (see later discussion about multiplication not being repeated addition).

First of all let me tell you that for kids, it a joy to learn ANYTHING, and they don't care how useful it may be later. What they care about is attention and praise and more attention and praise and **crucially**, *attention and praise.* Let me make that clear once more: what motivates children is *attention and praise.*[i]

Even when they are in their late teens, the subjects that they don't like are the ones that they don't think that they are good at. Why do they think that? Because they have had negative feedback (especially low grades). In other words, not enough *praise*.

The purpose of learning the times tables is to **avoid getting slowed down** by the arithmetic when, later, our loved one tries to *understand* relationships through the language *of* mathematics. And let me emphasise again, mathematics is a *language*.[ii]

As I mentioned earlier,[iii] one of the saddest findings in my years of research into the those factors that affect students'[iv] perceptions of their subjects, mathematics in particular, is that when we asked secondary school students, "What is mathematics about?" most of them said, it's about 'numbers'. This shows the real depth of

---

[i] And also curiosity, but that aspect of it is tangential to the immediate discussion.

[ii] 'Language' is defined as, "a complex system of communication"

[iii] But I don't mind repeating myself, especially since I am advocating repetition as a learning tool, but also because I want to put this into a broader context, so bear with me.

[iv] In the United Kingdom

misunderstanding that is rife about what mathematics really is. It is a tragedy. If we were to ask those same students, "What is English about?" or, "What is language about?", how many do you think, would say, "It's about the letters of the alphabet?"

And it is my contention that the tragedy starts with the times tables. This is because we falsely believe that children should only be introduced to the times-tables *after* they have learnt how to add.[i] I am here to stick my neck out and say, emphatically, that

THIS IS THE MISTAKE THAT IS HOLDING BACK OUR EDUCATIONAL STANDARDS IN MATHEMATICS.

Just as children need to memorise the shapes of letters and their associated sounds, without *understanding their meaning*,[19] they need to memorise their times tables without needing to know what it means. Similarly, would anyone advocate that children should know about Latin and Greek prefixes and the concept of angles before being taught that shapes of a certain category are called *tri*angles?

At this point, the academic in me is struggling to make itself heard. So, let's hear what the academic has to say.

Some academics have been trying for a long time to point out that knowledge cannot be separated from

---

[i] I'd even go as far as to say that children don't even need to know how to count in order to memorise their times tables.

understanding. This is in reaction to policy-makers' and practitioners' emphasis on "understanding"[i] as opposed to 'mere' "knowledge acquisition".[ii]

For example, Dr. Wu[iii] states:

> *"Facts vs. higher order thinking" is another example of a false choice that we often encounter these days, as if thinking of any sort—high or low—could exist outside of content knowledge. In mathematics education, this debate takes the form of "basic skills or conceptual understanding." This bogus dichotomy would seem to arise from a common misconception of mathematics held by a segment of the public and the education community: that the demand for precision and fluency in the execution of basic skills in school mathematics runs counter to the acquisition of conceptual understanding. The truth is that in mathematics, skills and understanding are completely intertwined. In most cases, the* **precision and fluency in the execution of the skills are the requisite vehicles to convey the conceptual understanding.** *There is not 'conceptual understanding' and 'problem-solving skill' on the one hand and 'basic skills' on the other. Nor can one acquire the former without the latter."*

At first glance, it may appear that there is a contradiction between what I am saying, i.e. that

---

[i] or higher order thinking as it is sometimes called

[ii] Sometimes falsely referred to as 'rote learning', a term that has become somewhat derogatory and is used to denigrate any kind of 'learning without understanding'.

[iii] Wu, H. (1999). Basic skills versus conceptual understanding. *American Educator*, *23*(3), 14-19. [https:// math.berkeley.edu/~wu/wu1999.pdf]

knowledge and understanding are separate entities and what Dr. Wu is saying, which is that the two are inseparable. However, on closer inspection,[i] we can see that the two arguments are, in fact, congruent. Let me explain.

To understand something, we need to have knowledge of its existence. This is what Dr. Wu calls, "the prerequisite vehicles". In other words, we need to have the vehicles **FIRST**. Therefore, it follows that we cannot expect people to understand something unless they have some knowledge of it and, in that sense, they are inseparable. However, it is also a good argument for saying - to those involved with education - that, students need to have the basic facts *before* they can try to make sense of them (develop an understanding). But the tone of Dr. Wu's argument could be interpreted to mean that mere knowledge is accompanied, *simultaneously* by understanding.[ii] I don't believe that this is what he means. My take on Dr. Wu's argument is that we can't expect understanding without *prior* exposure to the relevant facts. The tone of his argument is a reaction to the notion that factual knowledge is somehow inferior to understanding, which doesn't make sense, because

---

[i] Such as the part that I have highlighted in bold.

[ii] Although, going back to my discussion of the misinterpretation of Jean Piaget's work (see 'Notes to Parents' section), it is not that children will not understand at all what they *merely* memorise, it's just that the way they make sense of it is *qualitatively different* from the way they would do at a different stage.

factual knowledge is a *prerequisite* for understanding. I believe that Dr Wu's argument is a practical, not a philosophical, one and that what he is saying to educators is that we should value mere knowledge acquisition and not think of it as a poor substitute for understanding since understanding cannot happen without the acquisition of the necessary facts.

My argument here is that whilst it is obvious[i] that facts are required for understanding and that our minds try to make some sort of sense of any fact[ii] that enters them, children do not need to understand something, the way we intend for them to understand it later, in order to commit them to memory and to be able to recall them on demand. To illustrate, what is the missing word: Hickory, Dickory, _____. Did you need to have any conceptual understanding of each of those words in order to be able to recall the missing word? QED.

OK, that's enough from the academic, let's hear from the 'educator'[iii] again.

---

[i] At least to Dr. Wu and I
[ii] Or, indeed, fiction
[iii] Practitioner (?)

2
Memorising the Times Tables is NOT about Understanding Multiplication

Some people say that it is rote learning that is preventing some countries from progressing as fast as the more developed ones.[i] To me, that is like saying, it is because of food that obesity is becoming such a big problem in the US. It is not the tool, but how it is used that is the problem. It is when you confuse the times tables with the concept of multiplication that you lose sight of the purpose of the times tables. The times tables are like words in a language, you don't need to understand their etymology [ii] to learn and use them.[iii] Similarly, the child does not need to understand the concept of multiplication in order to be able to memorise, or indeed use, the times tables.

Let me tell you an anecdote that will, hopefully, illustrate the difference between learning the times tables and understanding multiplication. I will then explain why it is important to make such a distinction.

A few years ago, when I had just started doing research in psychology of education, I had the privilege of attending a talk by Professor Anna Sfard[iv]

---

[i] See http://en.wikipedia.org/wiki/Rote_learning#Development

[ii] The study of the roots of words

[iii] Although *after* we learn to use words, our understanding can be enriched by etymological exploration.

[iv] Professor of Mathematics Education

at the Institute of Education, University of London. She showed us a film where two four-year olds were playing with marbles that were in two separate boxes. At one point, each box contained four marbles.[i] When the children were asked if the number of marbles in each box were the same, they said, "No".

Although this was interesting, I didn't really understand the significance of it until about four years later when I started to teach a fifteen-year old girl who was struggling with her mathematics. She was able to add, subtract and multiply and also knew her times tables well, but she did not understand the concept of number.

You might be wondering how it could be that someone can multiply, add and subtract without understanding the concept of number; and this is exactly my point: The child does not need to understand what she is doing in order to be able to do the stuff.[ii] Then you might say, "But, wouldn't it be better if they did?" And I am here to argue this: NOT ALWAYS! AND NOT FOR THE TIMES TABLES and my mission here is to explain why, but bear with me for now.

When I began to suspect that there might be a discrepancy between my student's ability to

---

[i] Or it may have been a different number of marbles.

[ii] Or as we academics might say, to be able to carry out operations and manipulations

manipulate numbers and her ability to *understand* number, I went to our next lesson armed with an abacus. It was the kind of abacus that has ten horizontal rungs. On each of the top five rungs there were five green and five yellow beads and on each of the bottom five rungs there were five red and five blue beads, arranged to make four blocks of 25 beads.

I asked my student, "What is five plus five?" and she immediately said, "Ten". I then went to the abacus and on the first row of the beads, I separated the five green from the five yellow and said, "How many green beads are there on the first row?" She counted them, one by one, and *then* said, "Five." I then asked, "How many yellow beads are there?" She proceeded to count them, one by one, before telling me that there were five. I then asked her whether there were the same number of beads on each side of the rung and she said, "No." I then pushed the green beads back so that the green and yellow beads formed a continuous row again and then asked her, "How many beads are there on the top row now?" She proceeded to count them again, one by one, before telling me that there were ten. My student's mother who was witnessing this was showing very clear signs of distress that her daughter was apparently unable to answer such simple questions.

I had been thinking about this all the way through the session and suddenly, I remembered Anna Sfard's talk and the video of the two four-year olds who had said that the four marbles in the two boxes were not the same. Something in me seemed to click. I turned to the girl's parents and told them about what I had seen with the marbles and explained to them that what their daughter was doing was much more logical and 'normal' than what the rest of us do. She was looking at the yellow and green beads and could see that they were NOT the same, neither in colour nor their position on the rung. What we were expecting her to do was to take some abstract idea about the *number* of beads and to forget about all the other distinguishing features of those beads and to then declare that they were *the same*. How absurd is that?

This was, of course, the same logical thinking process that the four-year olds had been engaging in with the marbles. You could say that, unlike the rest of us, they had not yet lost their marbles [20]. Yes, I know they had been asked about the *number* of marbles, but they had not understood that, so they ignored it – something that we all do subconsciously quite often. If you don't believe me, when was the last time that you learnt a new word and noticed that it is being used more often than it had been before you learnt about it? Or bought a car and noticed that suddenly, there were many more of them on the streets than there were before you bought yours? This is related to the confirmation bias that I mentioned earlier.[i]

But, it was what happened next that really surprised me. From the following lesson onwards, my student began to dissociate the objects from the number that represented their quantity and began to *apply* her knowledge of the times tables to *calculate* the number of beads.[ii] For example, previously, when I had split the one hundred beads into two blocks of fifty and had asked her how many there were on each side, she had counted them, all fifty, one by one. Then, when I had moved one bead from one side to the other and had asked her again, "How many are there on each side now?" she had counted them all again, *one by one*. However, *after*[iii] my conversation with her *parents*, she started to tell me how many there were (in my example, 49 and 51) *without counting*.

You know how, sometimes, we don't get a joke until it is explained to us and then we find it funny? Well, it seems that the same thing had happened with this girl (for about ten years, unfortunately). No one had made the distinction for her, between the abstract idea of number and the reality of the objects. But when it was made *explicit*, she was able to extract the

---

[i] The late Douglas Adams (the author of 'The Hitchhiker's Guide to the Galaxy') called this SEP; "Somebody Else's Problem". You can read about it on Wikipedia.

[ii] Yep! She lost her marbles too ☺

[iii] That is, in the following session a few days later

abstract idea of number from the concrete objects that were the beads (or anything else)[21]. This came as a welcome surprise and an eye-opener for me because I was explaining the situation to her parents (and facing them) without being aware of the illuminative effect that this would have on my attentive student.[22]

Before this transformation, this girl might have been labelled with having 'dyscalculia'[i] but, at least in this case, it appears that her difficulty was not in her *ability* to understand, but her lack of *awareness* of the necessity to *dissociate*.[ii] Once this distinction was made for her, she was able understand it. Another way of looking at it is that she had not *associated* the operations that she was carrying out on paper[iii] with the (abstracted) reality of the number of objects around her.

I recounted this experience for you to make the point that, since my student was able to add, subtract and multiply without understanding the concept of number, it is possible for there to be NO relationship between *knowing* how to add, subtract or multiply and *understanding* the concepts of addition, subtraction and multiplication. Since these two are separate mental processes that work together, there is no reason for us to insist that children should

---

[i] Difficulty in understanding numbers

[ii] The number of objects from the objects themselves

[iii] Addition, subtraction, multiplication and division

understand the concept of multiplication before they learn their times tables by rote, or through other, possibly better, memorisation techniques (see later).

The trouble is that if we leave the memorisation process to a time when the child can *understand* the abstract concepts in mathematics, he or she will find it more difficult to memorise because the processes that come into play in trying to understand the ideas begin to get in the way of uninterrupted memorisation and subsequent speed of recall.

Another problem is that it often takes quite a long time for children to get to a stage [i] where they can really understand the idea of multiplication, so in assuming that they should understand at least *something* before they memorise their times tables, we begin to offer simplified *it'll-do-for-now* explanations, ignoring the fact that *it won't do for later*.

For example, we offer the 'simplified' explanation that multiplication is repeated addition because then we think that the children will have some idea of [ii] what they are doing when they are learning their times tables. But here is the glitch: Multiplication is NOT repeated addition. Look, if multiplication were repeated addition, multiplying a length, let's say two

---

[i] see my discussion on Piaget, later

[ii] And we call this understanding

metres, by another length, 'say three metres should give you six metres (another length), but this is not true. When we multiply two metres by three metres, we get six metres *squared*, which is an *area*, not a length; an area and a length are not the same thing.[i]

Do you see what I mean now? In insisting that children should understand what we teach them, we have to simplify, thinking that we are making it easier for them to understand when, in fact, we are distorting reality. And since, in our tests and exams, we only expect children to express their understanding in the distorted ways that we have taught them, those students who get to a point where they *do* understand that this does not make sense, begin to resent the 'lies' that they have been told and become disillusioned with the system. I have several bright young students in mind, as cases in point, as I write this. I had to have some serious talks with them before they were willing to engage with the system again.

There are many songs, poems and lines in plays that I memorised in my childhood without understanding their meaning until many years later. Had I waited until I could understand them, the chances are that I wouldn't be able to remember them as well now; not to mention all the opportunities for interacting with

---

[i] Now, we really *are* losing our marbles

my peers and learning a vast number of other things [i] in the process that I would have missed, including the opportunity to recall what I had memorised with a view to sharing them so that my associates could help me make better sense of them.

The same applies to the times tables. I remember having been taught my times tables as a song when I was in primary school in Iran in the late sixties/early seventies and, even though I have spent the last forty years of my life thinking predominantly in English, whenever I do multiplication, I refer to the song in Persian and then translate the answer back into English. Such is the power of memorisation at a young age. On the other hand, for those things that I did not learn by repetition, I usually have to go through the longer process by which I learnt them to remember them, which leads to delayed recall, because our minds code not just *what* we learn, but also *how* we learn it.[ii]

---

[i] "Things?" Yes, that's the technical term for them ☺

[ii] And it's the latter that facilitates *recall* and for the times tables, it's the efficiency of recall that is important.

3
Familiarity Brings Confidence

Confidence is a state of mind where doubt and fear are absent. One of our major sources of fear (and doubt) is the unknown. The antithesis to the unknown is familiarity, which is facilitated through repetition (frequency).

The more frequently we succeed, the more confident we become. The more frequently we are praised, the more confident we become. The more often we repeat a task, the more confident we become. The more confident we become, the more risks we are willing to take. The more risks we are willing to take, the more we experience failure. The more familiar we become with failure, the less anxious it makes us. The less anxious we feel about the prospect of failure, the more we are willing to move to the edge of our comfort zone, and venture beyond. The more we operate just beyond the edge of our comfort zone, the more we grow and the more empowered we become.

With regards to mathematics, the above spiral can begin with *familiarity* with the times tables, irrespective of the extent to which the underlying principles are 'understood'. The ensuing confidence to explore the domain of mathematics will lead to self-directed moves towards understanding.[i]

---

[i] Oops. It seems that I was off-guard, and the academic in me threw that bit in.

4
Memorising and Understanding Require Different Learning Techniques

Now that I have distinguished between memorising the times tables and understanding multiplication, I hope you will appreciate why the purpose of this book is not to teach multiplication as a concept, but to help you to help your children to memorise and most importantly, recall, their multiplication *facts* quickly and easily. Why is this important? Because **the *way* we teach should reflect our *reasons* for teaching.**

Imagine that you have a spanner in your house. It is a tool. You don't need to know how to use it to have it. Now imagine that you come across a situation where you need to use the spanner. If you did not have the spanner, you would need to call upon someone who has a spanner *and* knows how to use it to get the job done. Now consider what happens now that you have the spanner *of your own* at hand. You could take it out of the toolbox and experiment with it until you begin to understand how it works or if you aren't inclined to do that, you could ask someone who knows to show *you* how to use it rather than asking them to take *your* spanner to do the job, because it is *yours* and so *you* should know how to use it.[23] And even if you did not ask the person to show you how to use it, you are much more likely to pay close attention to what they do with *your* spanner. Also, the chances are that once you know how to use a spanner, you will start noticing situations where you can apply[i] your new-found skill.

Similarly, once children acquire the skill to recall their times tables quickly, they will look for, or notice, opportunities to apply it[ii] and, over time, through confident application of their knowledge of times tables to different situations, our children's understanding of the concept of multiplication will gradually improve, providing a solid basis for understanding more advanced mathematical ideas.

---

[i] And consequently, embed

[ii] And that's the start of a much more confident journey through mathematical ideas.

5
Memorisation of Facts is Not the Same as Rote Learning

Another potential source of confusion arises from the distinction between memorisation and rote learning. There are two ways of increasing the likelihood that we will remember something; repetition and association. Usually, the two work together. For example, if you see someone in the street, just once, it is unlikely that you will remember that person. If you see the same person twice, once by the beach in another city and once in the street somewhere else, you are still very unlikely to remember that person. However, if you see the same person in the same location more than once, you are much more likely to remember that person if you see him in the same location a third time. However, having seen the person twice in, let's say a particular restaurant, you are unlikely to recognise the person if you see him in the street on the other side of town; unless a third factor is activated.

Association of objects with objects, such as places to faces is a weak association and requires several repetitions for a rapid recall mechanism to be established. On the other hand, an object associated with an *emotion* is much more memorable, requiring much fewer repetitions for it to trigger recall. For example, music is emotive; it can change our mood. As such, it can trigger memories and acts as a very effective anchor for facilitating recall.

Rote learning is based on sheer repetition. Unlike the more natural way of remembering; by a combination of association, repetition and emotion, in rote learning, there is no attempt to associate (anchor) the object (or the concept) with anything else that could facilitate recall.[i]

Therefore, strictly speaking, learning with music and rhyme and movement, even though the music, the rhyme and the movement do not enhance *understanding* of what is being memorised, cannot be regarded as rote learning because the process does not rely on sheer repetition because the target material is being associated with specific melodies, rhymes or movements and, of course, emotions.

Therefore, the objective of acquiring just the facts, (without understanding at this stage, that is, simple memorisation and recall) can be achieved in a number of ways; rote learning being just one method. A more reliable approach, however, is through the use of anchors, that is associations, including music, body movements and, of course, mnemonics (see later).

---

[i] It's not quite as simple as that, but let's not get too distracted from the topic.

6
Memorisation is Not the Same as Recall

Have you ever been in a situation where you know that you know something, but you just can't remember it? What you are trying to remember is in your head, but there seems to be a problem *accessing* the information? This is the distinction between memory and recall.

It is important to make this distinction because how quickly we can recall what we learn depends on *how* we learn it. For example, visual memory is faster than auditory memory.[i]

Here is an example of two auditory mnemonics that have been devised to facilitate the recall of the order of the colours of the rainbow:

"ROY G BIV"
"Richard of York Gave Battles in Vain."

Which one do you think is faster? Since ROY G BIV is meaningless, it is harder to remember but, once remembered, it is faster to recall. On the other hand, "Richard of York Gave Battles in Vain" is easier to remember, in part because it can be visualised. However, since the pictures need to be deciphered into words and the words need to be subvocalized for

---

[i] A picture paints a thousand words (?). I put the question mark there because a word can also paint a thousand pictures in our minds ☺

the initial letters of each word to be decoded, it takes longer than the less meaningful ROY G BIV.

The point is that since the real purpose of learning the times tables is not to merely remember them, but to recall them as fast as possible, *on demand*, methods that we devise for learning the times tables should take factors that affect recall into consideration.

7
Younger Children Are Better at Memorising Without the Need to Understand

According to Jean Piaget, the father of modern child psychology, we go through 'stages of cognitive development'. In other words, we think differently as we grow. This discussion is important to learning the times tables because younger children are in the 'concrete stage' which means that they cannot understand abstract ideas yet. For example, the idea of five apples [concrete][24] can be split into two separate ideas, one apple [concrete] multiplied by five [abstract]. But you see, children in the 'concrete' stage are able, and happy, to memorise [25] abstract ideas, such as their times tables, without understanding them; as long as it has a payoff for them. Such payoffs could include attention,[i] praise or a reward.[ii]

This idea of stages would lead us to the conclusion that the more we want the child to understand the meaning of multiplication, the longer we have to wait. However, the longer we wait, the more opportunities we lose for children to memorise, with minimal interference from processes that try to 'understand'. In other words, the older children get, the more they

---

[i] which they not only understand, but value greatly
[ii] which is, itself, a more concrete form of attention and praise

try to understand[i] what they are doing, making it progressively more difficult for them to simply memorise what doesn't make immediate sense to them.[ii]

A few years ago, I was called to teach chemistry to a fifteen-year old student who was excelling at every subject except the sciences and mathematics. He was clearly an able student and so I needed to find out how he was looking at the subjects to make him not engage with science and maths. So, when he said to me, "What's the point of knowing how many electrons spin around a sodium atom?", I knew where the problem was. I said to him, "You probably don't remember this, but when you were about 5 years old, you were told to draw a line like this, (/) then draw a line like this (\) next to it, then draw a line like this (-) through them and say, 'A', and when you did, they would clap for you and say hurray and you really got a buzz from the reaction so you kept doing what they told you with the rest of your alphabet. At the time, you would not have been able to understand any explanations about how learning your alphabet would open a whole world of literature, text messaging, love letters and danger signs; you just trusted that it must be something useful [26] otherwise people who care

---

[i] Create meaning

[ii] Although they will still do it if it brings fringe benefits, such as praise or rewards.

about you wouldn't get so excited about you being able to do these seemingly pointless things." [27]

I went on to explain, "The problem with how we teach chemistry is that we are asking you to learn the ABCs of chemistry at an age when you ask 'why' and, in just the same way that we could not explain the joys of literature to you when we teach you the ABCs, I cannot tell you *why* it is important to know how many electrons spin around a sodium atom. You just need to *trust*. What I *can* tell you is that I can choose to look at this piece of paper and treat it just as anyone else would treat a piece of paper. On the other hand, I can choose to look at it as a chemist and start to focus on it and its chemical properties. I can then be amazed by how it is that it is made up of only three elements and that these three elements are the same three elements that make starch, which is so different to the sugar that we put in our tea, which is made up of the same three elements.[i] I can tell you why the paper is white, why it is not brittle, why it absorbs water when it gets wet but doesn't dissolve in it and why it is possible to fold and to tear it. On the other hand, to get the good feelings that come from reading literature, you need to find the right book before you can start to appreciate it. But with knowledge of chemistry, you can summon that feeling of awe and wonder at any

---

[i] And even in the same proportions: Cellulose is made up of Carbon, Hydrogen and Oxygen in the same ratios (1:2:1) as starch and sucrose, but with such different properties ... Awesome!

time by choosing to focus deeply on *any* object that is around you, such as your very skin."

My pupil said something like, "Wow. No one had ever told us that this was just the ABCs, so here we were trying to make sense of the Periodic Table and couldn't understand it and decided that it was pointless." Two months later, he went on to achieve A grades in both Mathematics and Science at GCSE.

And that leads me nicely to the next section which is about how we take the understandings from these seven principles and apply them in a way that will maximise their impact. I have called them the seven Rs (think of them as seven 'reinforcers')

> "Don't ask yourself what the world needs. Ask yourself what makes you come alive and then go do that. Because what the world needs is people who have come alive."
> — Howard Thurman

> "I would rather discover one scientific fact than become King of Persia."
> — Democritus (460-370 B. C)

> "God does not care about our mathematical difficulties. He integrates empirically."
> — Albert Einstein

Seven Rs
(Reinforcers)

> "The belief that all genuine education comes about through experience does not mean that all experiences are genuinely or equally educative."
>
> - John Dewey

> "The danger of the past was that men became slaves. The danger of the future is that men may become robots."
>
> - Erich Fromm

> "Your task is not to seek for love, but merely to seek and find all the barriers within yourself that you have built against it."
>
> - Rumi

Application of the Seven Principles to the memorisation of the times tables

So far, I have argued that, in practice, it will be of great advantage for our children [i] to memorise their times tables as early as possible. I have argued that the purpose of doing this is **not** for them to understand the concept of multiplication. The purpose is for children to be able to **recall** their times tables with minimal interference from other mental processes. Utilising our current understanding of the how our memory works, it is possible to increase the effectiveness with which we do this.

Another important psychological factor is **motivation**. This means making it interesting/desirable. I have already touched on the importance of attention and praise. Beyond that, as children grow older, what is considered as 'interesting' changes. For younger children, it not so much 'meaning' that makes something interesting as 'interaction'. Younger children are more likely to be motivated by what can be phrased as, "I wonder what I can do with this". For older children, the question becomes, "I wonder what this can do for me."

Bearing in mind the principles that I mentioned earlier, in this section, I have summarised the factors that need to be borne in mind when teaching the times tables into what I have called Seven Rs.

---

[i] And society

1. Respect

Engagement, not enforcement.

We sometimes assume that very young children do not appreciate concepts, such as 'respect' and this needs to be taught to them. However, research[i] has shown that even 18-month old infants can understand the idea of 'intention'. I have personally seen children as young as 12 months old demonstrate such understanding. This means that they can understand the difference between, "Don't do that" when the underlying reason for saying it is, 'because it annoys *me*' and, "Don't do that" when the underlying reason is, 'because I am concerned about *your* welfare'.

Children are more likely to engage with us if they feel respected. In practice, this means that they need to feel that, in asking them to do anything, our focus is primarily on their welfare and their needs, not our own. It does not matter whether or not they understand what we are asking them to do, but they need to *feel* that our *intention* is directed at their welfare and not at our own convenience.

I remember when my daughter was less than two years old, she wanted to cross the road from between parked cars and did not show much interest in my

---

[i] E.g. Meltzoff, A. N. (1995). Understanding the intentions of others: re-enactment of intended acts by 18-month-old children. *Developmental psychology*, *31*(5), 838.

instructions that she should not attempt to cross the road by herself. I knelt down to her level and pointed out what I could see from there. I then picked her up and asked her to pay attention to the difference between what I could see from her eye level and my eye level. I then said to her that from my eye level I could look out for her safety in a way that she was not able to do *yet*. I noticed that after that, she was less resistant to my instructions. I could have achieved this by threatening to punish her or by rewarding her or by shocking her by raising my voice suddenly. However, showing respect for her intelligence and demonstrating that my instructions were intended to benefit her, and not myself, was not only effective, but also more sustainable.

The relevance of this to teaching the times tables is that to engage children, we need to respect their intelligence and needs. We respect their intelligence by not pretending that they would not understand if our words do not match our real intentions. And in terms of needs? *Attention and praise.* If every time that they see their times tables, they are reminded of the attention[i] and praise[ii] and fun[iii] that they had, then the motivation and engagement will already be there.

---

[i] Such as sitting on mummy's lap and exploring

[ii] Such as seeing people smiling at her for being able to repeat what she's told correctly.

[iii] Such as dancing, singing, exploring, playing peek-a-boo, etc.

2. Random

Earlier, I argued that in attempting to help children to understand the times tables we slow down their recall. Whenever I see resources for teaching the times tables, the numbers are presented in a form that allows children to spot patterns in them. For example, looking at the 2s column in the times tables allows the children to notice that it is counting up in twos. This means that the child will begin to try to remember *sequences*. **This slows down recall.**

Therefore, it is important to introduce the times tables facts to children in a way that does not allow for 'pattern spotting'. The exercises in this book take this into consideration by presenting the times tables facts in the form of **triads**. This has a number of other advantages that I will discuss later.

3. Rapid Recall

Since speed of recall is of the essence and since visual recall is faster than auditory recall and since memorisation is reinforced by association, these need to be taken into consideration both when designing methods to facilitate the memorisation of the times tables and when considering ways of implementing those methods.

The exercises described in this book make use of a combination of visual and auditory cues.

4. Rhythm and Melody

The addition of rhythm and melody would add an additional layer of association. Rhythm also engages the body, a further layer of association that facilitates memorisation and recall. As such, I encourage you to introduce rhythm and melody to each of the exercises where possible.

5. Relate (Parallel Learning)

Any combination of stimuli that our brain receives simultaneously, irrespective of the sensory modality is automatically linked.[i] If the child keeps on seeing the numbers 6, 8 and 48 appearing together, pretty soon, the child will be able to identify the missing number in a triad, when presented with only two of them. This is how the exercises are set out.

In addition, the triads are presented in a 'triangle' format with the product of the two numbers being on top. This arrangement has the added advantage that it makes the transition to division a spontaneous one whereby the child will not need to memorise the times tables backwards in order to be able to divide.[28]

---

[i] Think Ivan Pavlov and his dogs.

6. Reward

Having reward charts where the child can get stickers for being able to compete each triad is a good way of engaging children. Remember that a reward is a tangible (concrete) form of attention and praise that children can relate to easily. It also has the advantage of being a visual anchor that will encourage them to want to get more attention and praise and so will be more enthusiastic to do it again and again. Which links neatly with the last R.

7. Repetition

It's a cliché but it is true; repetition is the mother of skill. The exercises are designed in such a way that the child will be exposed to the same triads four times in every learning cycle without appearing repetitive.

The Wrong Ways of Teaching the Times Tables

Mathematics is not hard; mathematics is boring when it is slow and laborious. It is slow and laborious when you don't have quick access to its main tool, the times tables. *Yes, the times tables is the key.* If you don't know it, it slows you down. And when something feels laborious, you don't do it with enthusiasm, and when you don't do something with enthusiasm, you eventually fall behind and the more you fall behind others[i] the more it affects your self-concept. And all because we slowed down the memorisation and recall process by insisting that the process has to have an element of 'meaning', albeit just the recognition of 'patterns'. As I have argued above, trying to impart any kind of meaning slows down [29] the process of memorisation and recall, and, *"that is not helpful"*.[ii]

When trying to help children to *memorise* their times tables, avoid resources that try to teach the *concept* of multiplication. Most often what I see is either a Times Tables Square/Grid or a series of multiplication sequences (shown on the next page).

Most parents and teachers who have tried these methods will tell you that children tend to use these in one of two ways. Either as a reference (the look-up method) or for sequencing (the count-up method).

---

[i] or more to the point, expectations. Actually, standardisations and comparisons which induce fallacious competition are amongst my other bones of contention with the education system, but let's leave those for another day.

[ii] As my eight-year old son is fond of saying.

The multiplication square or grid

×	1	2	3	4	5	6	7	8	9	10
1	1	2	3	4	5	6	7	8	9	10
2	2	4	6	8	10	12	14	16	18	20
3	3	6	9	12	15	18	21	24	27	30
4	4	8	12	16	20	24	28	32	36	40
5	5	10	15	20	25	30	35	40	45	50
6	6	12	18	24	30	36	42	48	54	60
7	7	14	21	28	35	42	49	56	63	70
8	8	16	24	32	40	48	56	64	72	80
9	9	18	27	36	45	54	63	72	81	90
10	10	20	30	40	50	60	70	80	90	100

Multiplication sequences

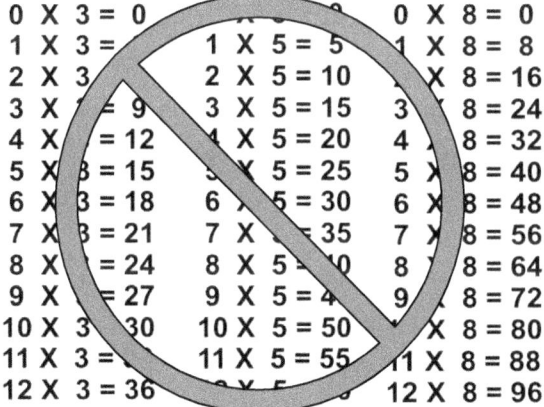

```
0 X 3 = 0        0 X 5 = 0        0 X 8 = 0
1 X 3 = 3        1 X 5 = 5        1 X 8 = 8
2 X 3 = 6        2 X 5 = 10       2 X 8 = 16
3 X 3 = 9        3 X 5 = 15       3 X 8 = 24
4 X 3 = 12       4 X 5 = 20       4 X 8 = 32
5 X 3 = 15       5 X 5 = 25       5 X 8 = 40
6 X 3 = 18       6 X 5 = 30       6 X 8 = 48
7 X 3 = 21       7 X 5 = 35       7 X 8 = 56
8 X 3 = 24       8 X 5 = 40       8 X 8 = 64
9 X 3 = 27       9 X 5 = 45       9 X 8 = 72
10 X 3 = 30      10 X 5 = 50      10 X 8 = 80
11 X 3 = 33      11 X 5 = 55      11 X 8 = 88
12 X 3 = 36      12 X 5 = 60      12 X 8 = 96
```

'Pattern Seeking' Methods

Some people get understandably excited when they find patterns in the times tables. Unfortunately they falsely believe that using these patterns to *derive* answers to basic times tables is a good way of 'learning' them.

For example, why bother learning that nine eights are seventy two when, "Well, you know that when you multiply any number by nine you can, erm ... take one away from that number, then find the difference between that number and nine and then put that number after the first." Yes, it works, but if you want to do that, just use a calculator!

Your budding mathematics genius needs to be able to see the numbers 8, 9 and 72 almost simultaneously or instantaneously in their head as a *trio* that 'hang around together'. Otherwise, she's going to have problems when it comes to dividing 72 by 8. The 'working it out method' doesn't work for division.

Another 'pattern-based' method I saw being suggested with pride on the internet, is for the eight times tables. The suggestion is this:

Write these numbers down in order: 0123445678
then write these numbers next to them: 8642086420

Notice that the first row is just like counting - except that you have to remember to put the extra four in - and the second row is counting backwards in twos, starting from eight and repeating when you get to zero.

In spite his effort and enthusiasm, such a student is prone to suffer the fate of the Alex in the Introduction.

Other Pattern-Based Methods

Some well-meaning enthusiasts advocate the use of pictures where items, let's say strawberries, are grouped together. For example, we could have four boxes each containing six strawberries. These are designed to explain to the children how the number twenty four can be derived from four and six.

The problem with this method is that, being children, they will be more interested in the strawberries and their boxes and their colour and shape and size and taste and the like rather than the relationship between the total number and the way in which they are grouped. Even if we do manage to focus their attention on the abstract ideas of number, their clever little brains will be looking to spot ways of avoiding memorising. The best way to do this is to look for patterns. Once they notice that they can visualise a grid and count the squares (or imaginary strawberries), they[i] will not be convinced that memorisation is warranted. As such, we might as well not complicate the matter by introducing concrete objects in the first place.

---

[i] I mean their subconscious mind (difficult to convince through the intellect)

Story-Based Methods

Other well-meaning enthusiasts advocate the use of stories in order to make the concepts more concrete and more memorable. Using rhymes as mnemonics for numbers makes it easier to make up stories. These stories are then expected to help children to associate the three numbers.

For example, we can first create associations between objects and numbers in this way, one (bun), two (shoe), three (tree), four (door), ... eight (skate). We now have elements that can be turned into a story. Here's one that I just made up: When Jack opened up his birthday present, he threw the shoe (two) out of the door (four) and shouted, "I said, I wanted a skate (eight)."

There are many variations on this theme with illustrations to help make the connections. The problem with these methods is not in the memorisation part. Yes, it might make it more fun, 'meaningful' or even memorable, but trying to remember a 'story', simply to extract three numbers out of it, weakens the pure association between the three numbers and slows down *recall*.

Compare that with, 'Hickory, Dickory, Dock'. No meaning, no complex associations through characterisation of the elements, just three meaningless words. OK, they are followed by a rhyme, so we could try something like this: "Nine, eight, seventy two; that's all you need to do."[i] And if you say each hemistich in a 'cha cha cha' rhythm, you'll have a strong competitor for 'Hickory, Dickory, Dock.'

---

[i] Another little something that I just made up.

The 'Look-up' method

In the Look-up method,[i] the grid is used as reference for when the child needs to multiply two numbers. They will scan for one of the numbers along the top and then the other number along the side and check to see where the row and column meet. Fine, they have found the number, but as every experienced teacher and parent would tell you, children do not acquire quick recall of their times tables this way, no matter how often they do this.[30]

With these methods, students can see that in each row, the numbers increase by a certain amount and the same is true for each column and so, they will develop[ii] an understanding of what multiplication is. In fact they don't, because multiplication is not successive addition.[iii] This is one of those fallacies that we tell our children because we feel that they ought to "understand what they are doing when they are using their times tables." My many years of experience as a teacher, private tutor and parent has taught me that children resent being patronized; even if they find out much later that the half-truth that they were told was designed to make the concept easier to teach.[31]

---

[i] more often used in conjunction with the Times Table Grid

[ii] or so we think

[iii] As I've already pointed out

The 'Count-up' Method

Multiplication sequences lend themselves more readily to children memorising each sequence, such as 3,6,9,12,18, ... This method is even more problematic than the first because the child will 'count' (often with their fingers) as they recite the sequence that they have memorised. For example, if asked to multiply three by four, the child will say, whilst counting with his fingers, "three, six, nine, twelve" and then, he will look at his hand to check that he has closed 4 fingers and then he will look up, smile at you and proudly say, "twelve" and then we smile back at him and pat ourselves on our metaphorical backs thinking that we have done our child a favour.

NO, WE HAVE NOT, because the object of the exercise is to develop *rapid recall* NOT *to understand the process.* Of course it is not the sweet child's fault and so, of course we shouldn't penalize him for doing this. We should really be chastising ourselves for letting it come to this. But we should then smile sweetly at the child and let him know that *we are not interested in the answer if it takes longer than half-a-second to produce, even when it is right.* However, we can only do this if we have alternative ways of teaching the times tables.

AND THE REST OF THIS BOOK IS ABOUT ONE OF THOSE ALTERNATIVE WAYS.

The Triad Method [i]

---
[i] Patent pending

Background

The typical times tables grid is a jumble of numbers that does not lend itself to being memorised and if a child is encouraged to try to learn it, he will start to look for patterns that might make the task easier. I have already explained the problem with this.

	2	3	4	5	6	7	8	9	10
2	4	6	8	10	12	14	16	18	20
3	6	9	12	15	18	21	24	27	30
4	8	12	16	20	24	28	32	36	40
5	10	15	20	25	30	35	40	45	50
6	12	18	24	30	36	42	48	54	60
7	14	21	28	35	42	49	56	63	70
8	16	24	32	40	48	56	64	72	80
9	18	27	36	45	54	63	72	81	90
10	20	30	40	50	60	70	80	90	100

The alternative is to consider it a table that acts as a reference guide for when one needs to know the product of two numbers. Since one can always look these up, children - being the intelligent beings that they are - will not be convinced that memorising all those numbers is warranted.

However, if we consider that each multiplication is a group of three numbers, a triad, then there are only 36 groups of three numbers that need to be memorised.[i] This looks more manageable:

	2	3	4	5	6	7	8	9
2	4	6	8	10	12	14	16	18
3		9	12	15	18	21	24	27
4			16	20	24	28	32	36
5				25	30	35	40	45
6					36	42	48	54
7						49	56	63
8							64	72
9								81

However, such a grid still suffers from the drawbacks of "Pattern Spotting".

---

[i] I have removed the ten times tables because adding a zero to the end of the number need not be learnt through repetition

How the Triad Method ⁱ works

This method enables children to memorise each of the above 36 triads in a way that, not only facilitates instantaneous recall, but also allows them to deal with division *when* the time comes.

With this method, the child learns to associate sets of three numbers with each other as a single group. Each group is designed to create what is called a 'gestalt'.

To give you an idea of how a gestalt works, think about an island. Whenever you do this, three separate ideas of 'sea', 'land' and 'water' come together to form a single gestalt, in this case, an island. In other words, I can ask you to think about a sea and you can do so without invoking the image of an island. Similarly I can ask you to imagine 'land' and you can do this without thinking about any islands. Finally, I can ask you think about something surrounding something else, and you can do this without an island coming to mind. However, when I ask you to think of an island, you cannot do so without invoking the concepts of land, sea and surround, all at the same time.

What's even more interesting is that, in a gestalt, you not only remember the individual elements that need to go into creating the gestalt, but also the structure of the relationship between the elements. For example, the word 'lake', also invokes the same three concepts

---

[i] Patent pending.

of land, sea and surround, but the relationship is different so that if I ask you to think about a lake, the idea of an island does not interfere with it at all. In other words, you don't think, "Oh, that's confusing, how did those three ideas of land, sea and surround go together to make this one? Was it the land surrounding the sea or the other way around?" That does not happen because in a gestalt, the relationships between the elements (concepts) are just as important as the elements themselves.

Finally, in a gestalt, all the elements that make up the gestalt come to mind instantaneously. In other words, when we hear the word 'island', we do not imagine some land and then isolate it in some way before adding water to the region on the other side of the boundary, the whole thing is *one single concept*.

This is the idea behind the Triad Method, that is, learning to associate three numbers, such as, 7, 8 and 56 as a gestalt.

in NLP[i] circles, a 'gestalt' is called a 'chunk'. The advantage of calling it a 'chunk' is that it can also be used as a verb so that we can also give a name for the *process* by which this happens; *chunking*.[ii]

---

[i] Neuro-Linguistic Programming.

[ii] Whereas no one ever talks about 'gestalting'.

The Anatomy of a Multiplication Expression

In this section I explain why multiplication is not repeated addition. I also introduce some of the words that I will be using when explaining the exercises.

Just as in English, the order of words changes its meaning, such as 'outdoor play' and 'play outdoor', in the same is true for a multiplication expression. That means that which number comes first in a multiplication expression depends on what we are trying to say. For this reason, each of the two numbers that we multiply together (each of the 'factors') have different names depending on which one comes first (see below). This means that there is a difference between 3 x 7 = 35 and 7 x 3 = 35.

Consider the situation where you have 3 bags of marbles and there are 7 marbles in each bag. This can be represented mathematically as: 3 x 7 = 35.

This means: 3 bags x 7 marbles **per bag** = 35 marbles

Now imagine that you have seven bags, each containing three marbles.

This time, the equation would be written as: 7 x 3 = 35

That is, "7 bags x 3 marbles **per bag** = 35 marbles [i]

---

[i] Same result? Yes, but only if you are only interested in the destination and not the journey.

Now, consider a rectangle having a width of 5 cm and a height of 3 cm; it has 3 rows and 5 columns.

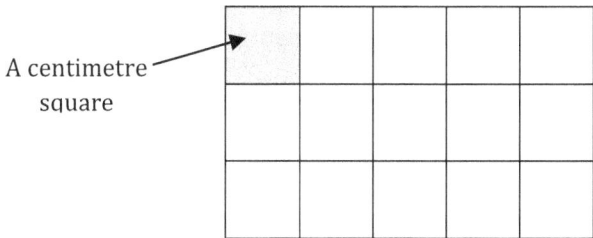

When asked what the area of the rectangle is, we can say,

$$3 \times 5 = 15 \quad \text{OR} \quad 5 \times 3 = 15$$

However, even though the result is the same, in the language of mathematics, these are different statements. The first statement says:

3 rows x 5 centimetre squares **per row** = 15 cm squares

The second statement is:

5 columns x 3 cm squares **per column** = 15 cm squares

We can write this as:

Multiplicand x Multiplier = Product

In practice, the multiplier can be represented as *something per something else*.

Finally, the multiplicand the multiplier are collectively called *factors*.[i]

Notice that we are not adding 3 cm five times or 5 cm three times, otherwise we would have 15 cm and not 15 cm *squares*

---

[i] Just as men and women are collectively called people.

> "My special pleasure in mathematics rested particularly on its purely speculative part."
>
> — Bernhard Bolzano

> "In every block of marble I see a statue as plain as though it stood before me, shaped and perfect in attitude and action. I have only to hew away the rough walls that imprison the lovely apparition to reveal it to the other eyes as mine see it."
>
> — Michelangelo

> "What you do today is important, because you are sacrificing a day of your life for it."
>
> — Anon

Example Exercises

Introduction

In these exercises, the purpose is to help children to perceive the numbers as being groups of three; that is, as 'triads', with, hopefully, each triad becoming a gestalt.

For smaller children, this can also be an exercise in number recognition. They don't even need to know how to count, or be able to read, before starting on this exercise. All they are required to do is recognize the patterns (which they are exceedingly good at).

 Just as children can associate the sound of the word 'spider' to a pattern that looks like this (without ever having seen a spider),

| 72 | they can just as easily associate the sound 'seventy two' to a pattern that looks like this (without needing to know what it means). |

There are numerous ways in which groups of items (in our case, three numbers) can be presented so that they become associated as a gestalt. Here, I have presented just one way. However, once we know the purpose of the exercise, we can devise new methods for different situations.[i]

Here, I have chosen to present the triads as a boat.

---

[i] We just need to be careful not to 'reinvent the wheel'.

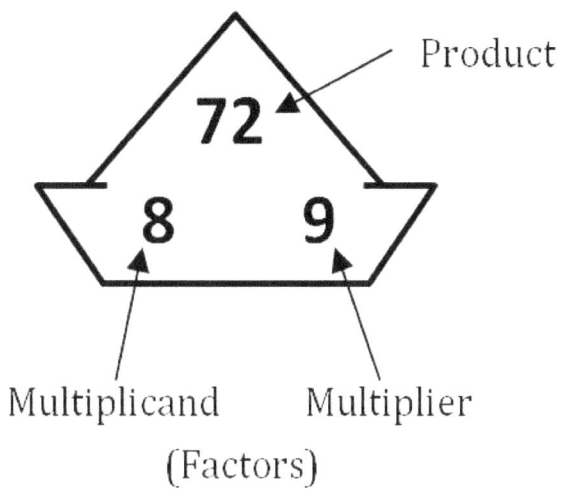

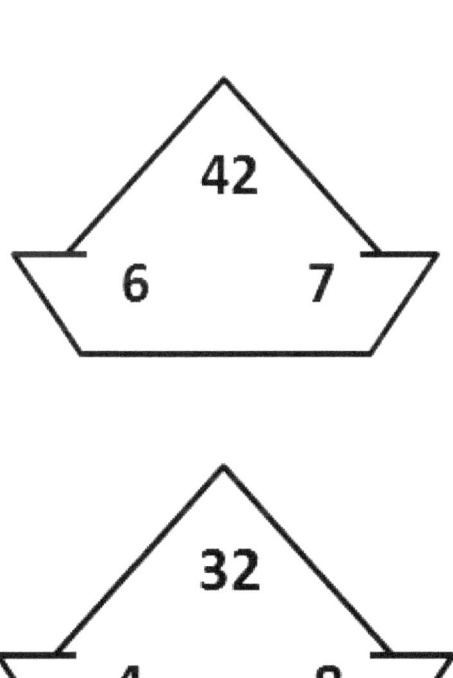

Why a boat?

One of the advantages of a boat shape for each triad is in that it creates a seamless link to division. Let me explain.

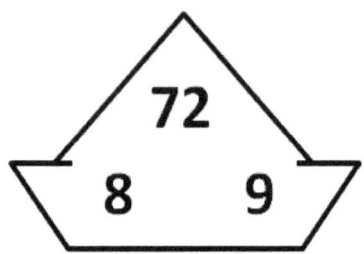

The way to read this to your child is:

"Eight, nine, seventy two" pointing to each number quickly (your finger should make a rapid and smooth anticlockwise circular movement, starting from the bottom left).

Note that there is no reason to say, eight *times* nine *is*[i] seventy two. If your child is visually dominant, it will make less difference, but if, like me, he is auditory dominant, every time he wants to access his times tables, in his head, he will repeat the process by which he learnt the triad. The extra sounds that he has to say in his head[ii] (*times, is* or *equals*), will just slow his recall down unnecessarily.

In the first stage, the child is repeatedly exposed to the triad. For example, "Eight, nine, seventy two".

---

[i] Or 'equals' (which takes even more time)

[ii] This is called subvocalisation

In the second stage, the product is obscured and the child will need to complete the phrase; "Eight, nine, _____ [pause for child to complete]".

In the third stage, the multiplier is obscured and the child will need to insert the missing number or sound[i] "Eight, _____ [pause for child to complete] , seventy two".

Finally, in the fourth stage, the multiplicand is obscured and the child will be asked to say the complete triad by inserting the missing number or *sounds*: "Hm? _____, nine, seventy two?".

Now, the magic is this. In stage two and three, obscuring any of the numbers in the hull of the boat, *automatically* reverses the multiplication.[32] In this example, if you put your finger on the nine, the result can be read as, "seventy two divided by eight is nine." Another way of saying 'divided by' is 'over'. For example, "seventy two *over* eight is nine". That's because a 'fraction' is another way of showing division and, when a division is expressed as a fraction, what would be the **product** of a multiplication is written **over** a **factor** of that multiplication and the result is the other factor. In this example,[ii]

$$72 \div 8 = \frac{72}{8} = 9 \qquad \text{Notice that,} \quad \div \longrightarrow \frac{72}{8}$$

---

[i] Or shape, if the pattern is completed by filling a blank on paper. Younger children don't even need to know that these represent a number.

[ii] Notice that the 'divide sign' that we introduce to children as the first symbol for division is a representation of a fraction.

Similarly, "seventy two *over* nine is eight". And this is what the boat shows.

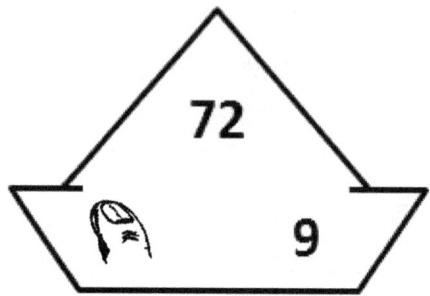

Having explained all that, I want to stress that, for this exercise, **the children are not learning about division**. In other words, they are not learning what division means; that comes much later. At this stage they are just **developing a gestalt** which gives them the ability to *rapidly* recall missing numbers from a times tables triad, **not why it works, or even its purpose**.

For the exercises that I have prepared based on this method, the triads are not in an order that one might expect from traditional times tables grids. This is deliberate. I take every precaution to minimize pattern spotting at this stage.[i] Each page contains more than one triad.[ii] This provides a challenge where the child learns that it is necessary to distinguish between different triads and that each must be remembered as a group of three that are separate.

---

[i] See above, the Seven Rs (Random).

[ii] I experimented with one triad per page and I found it to be less effective because the child's mind is not primed to be aware that it needs to distinguish between different triads. The lack of contrast makes each triad less memorable.

Stage 1
Introduce the triad

Starting with the multiplicand, move your finger anticlockwise in a smooth circular motion and say the numbers. For example, for the following triad, point to the relevant number and say, "Eight – nine - seventy two" in a rhythmic tone, such as the way you might say, 'Humpty - dumpty - sat on a wall'. Repeat[i] this several times, then go onto the next triad.

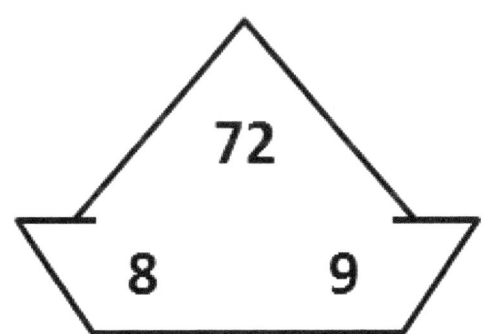

Eight, Nine, Seventy Two.

---

[i] See above, the 7 Rs (4. Repeat).

Stage 2
Hide the Product

Ask, "What's the hidden number?", then pause for one or two seconds. If the child does not reply, or responds incorrectly, then place your *other* finger under the multiplicand and, in the same rhythm as in Stage 1, read out the number. Then move to the multiplier and read out that number and then pause for a few seconds. If the child is not able to remember the third number in the triad, go back to Stage 1 for this triad and start again.

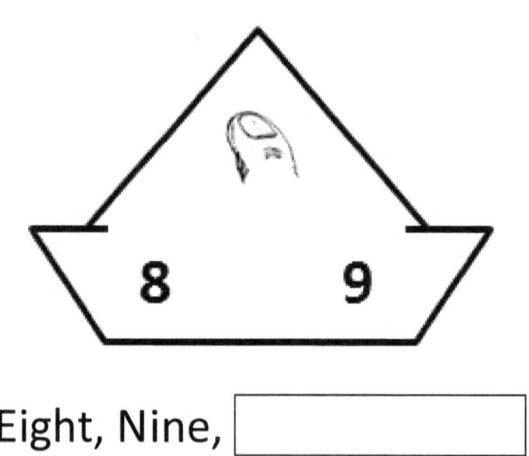

Eight, Nine,

Stage 3
Hide the Multiplier

Ask, "What's the hidden number?", then pause for one or two seconds. If the child does not reply, or responds incorrectly, then put your other finger under the multiplicand and say its name. Pause briefly, and then move onto the product and read it out – all in the same rhythm as in the 'introducing the triads' section. Again, if the child is not able to remember the number, go back to Stage 1 for this Triad and start again.

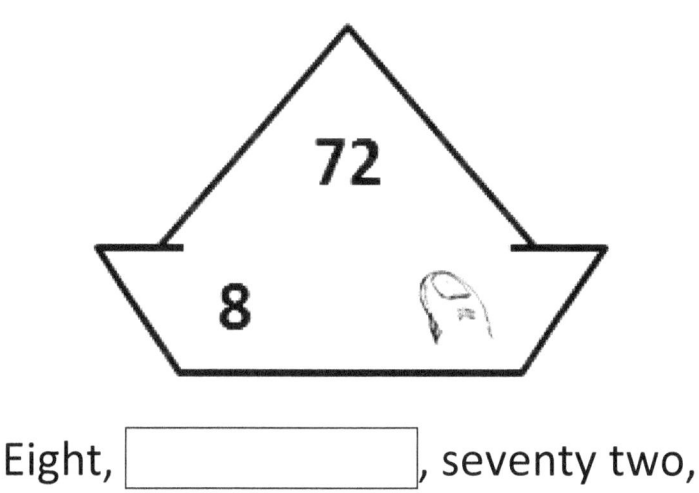

Eight, ⬚ , seventy two,

Stage 4
Hide the Multiplicand

Finally, move your finger over the multiplicand and ask, "What's the hidden number?", then pause for one or two seconds. If the child does not reply, or responds incorrectly, then read out the next two numbers in the same rhythm as in Stage 1, moving your other finger smoothly anticlockwise to each of the other two numbers. Again, if the child is not able to remember the number, go back to Stage 1 for this Triad and start again.

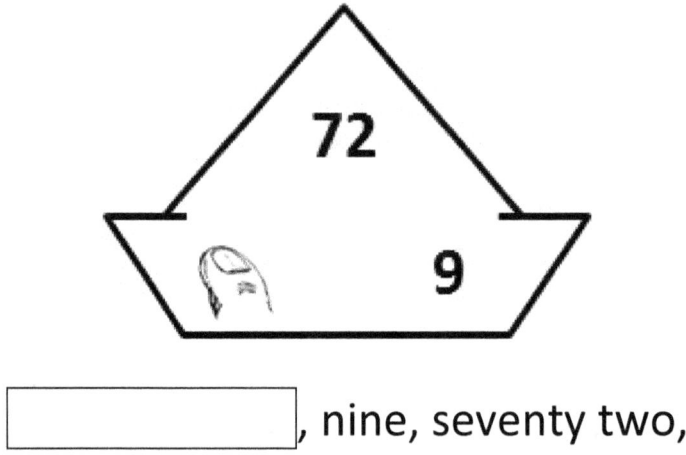

☐ , nine, seventy two,

The 36 Triads

On the following pages, you will find everything that your child needs to know to become comfortable with the times tables. On each sheet, there are three triads or boats.

After going through a set of three (one sheet), cover up one number from each triad and see whether your child can remember the missing numbers. If she can, she can get some attention and praise. If she can't, she can get to spend more time with the adult to learn.[i]

---

[i] How's that for a win-win situation for your child?

Set 1

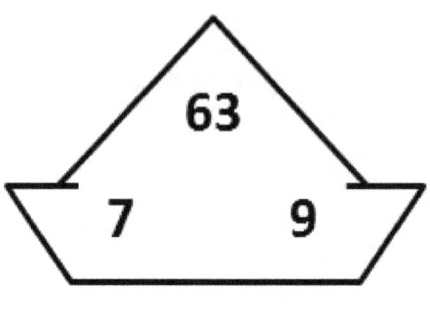

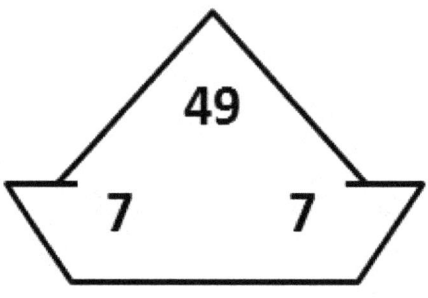

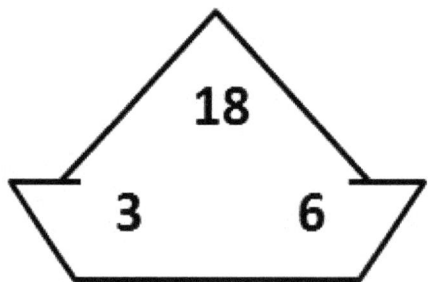

Set 2

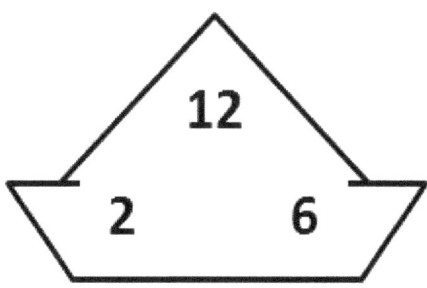

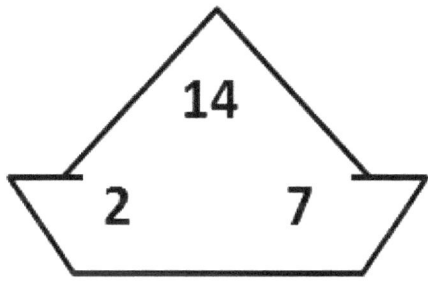

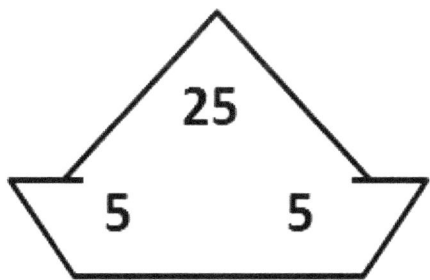

Set 3

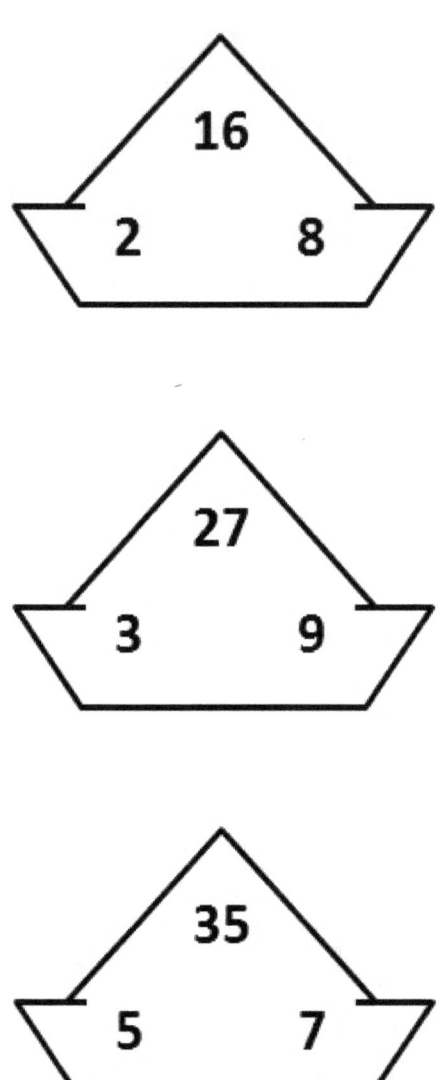

Set 4

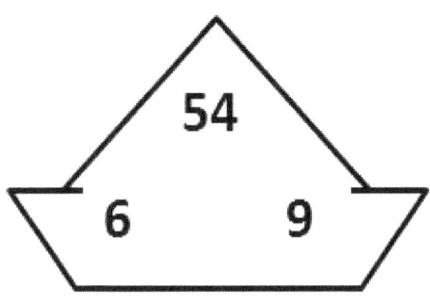

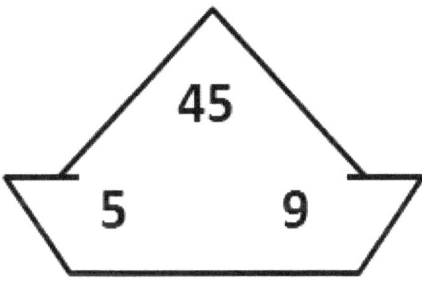

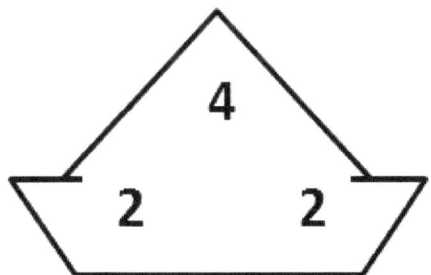

Set 5

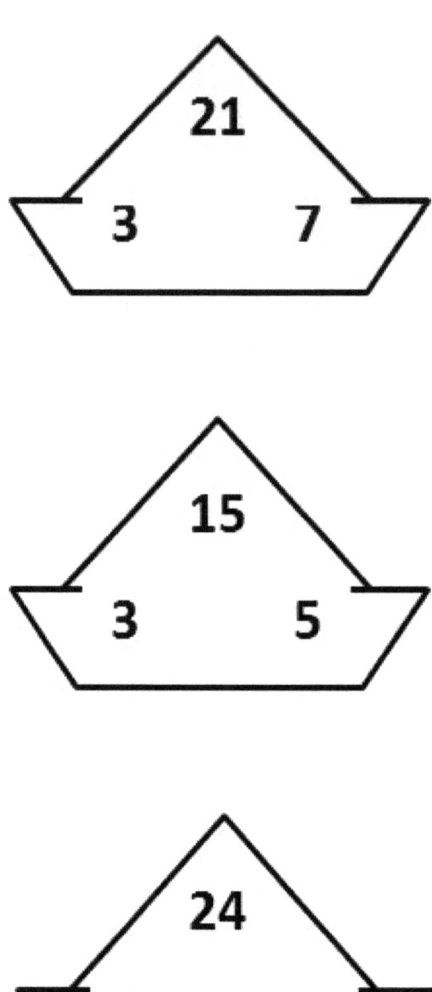

Set 6

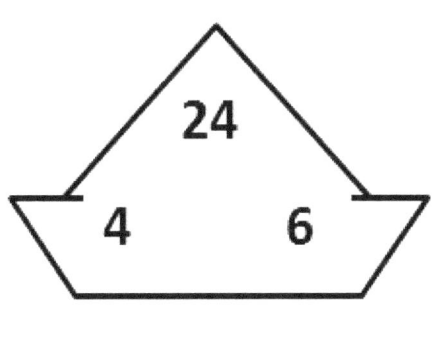

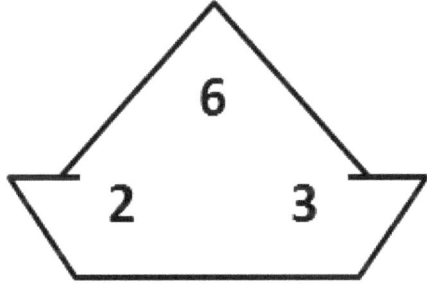

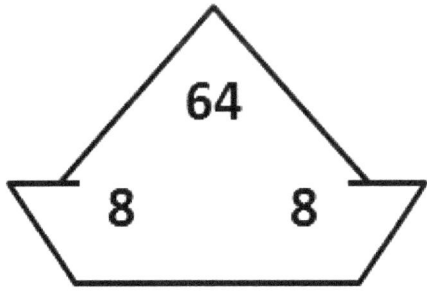

Set 7

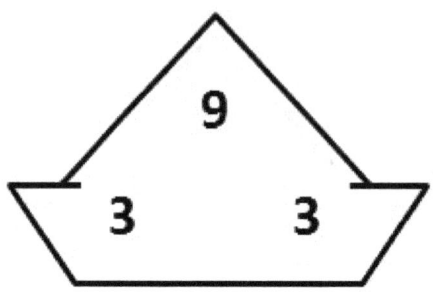

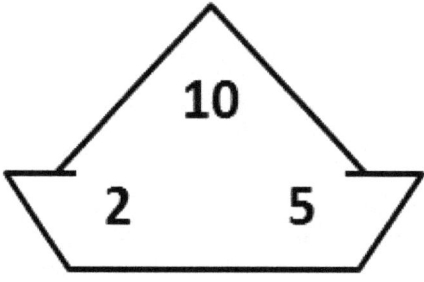

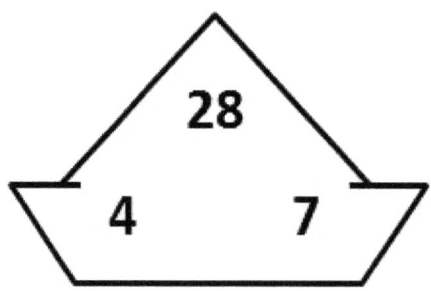

Set 8

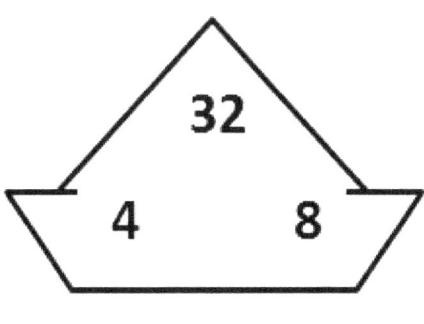

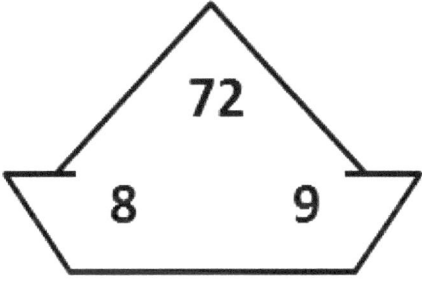

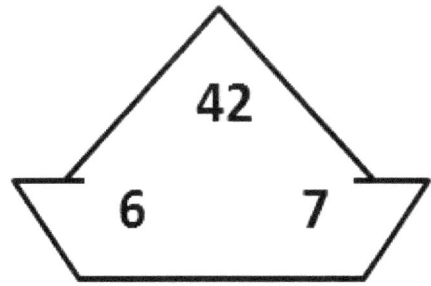

Set 9

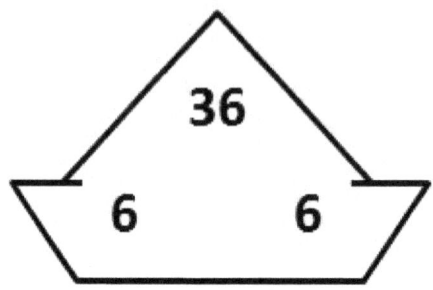

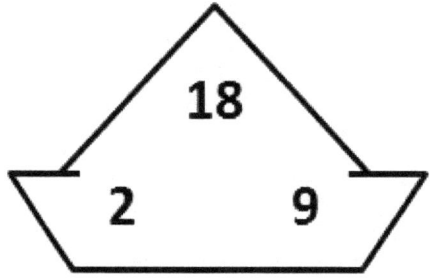

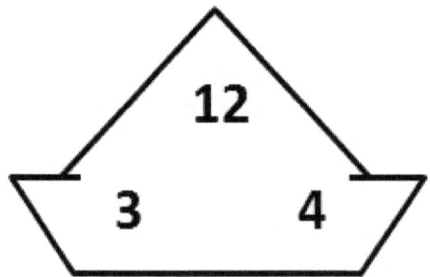

Set 10

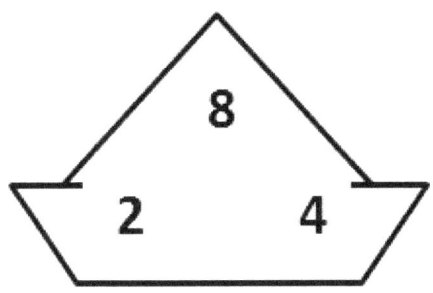

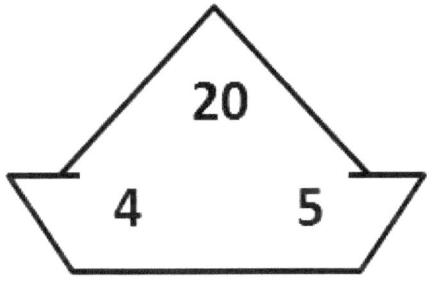

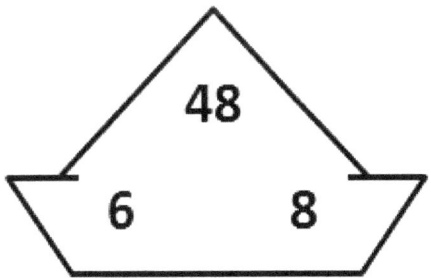

Set 11

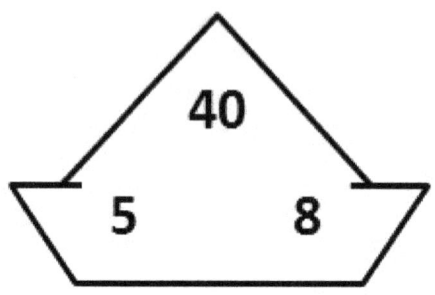

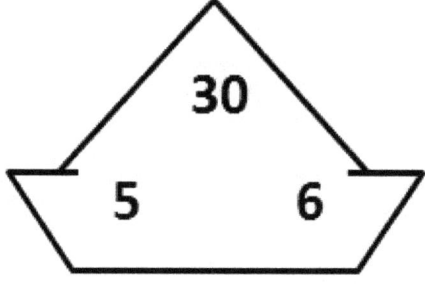

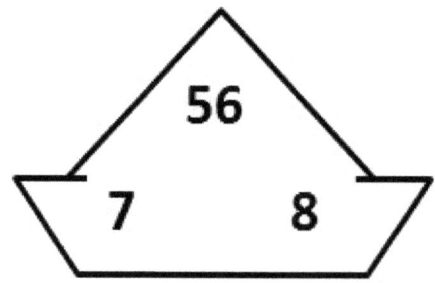

Set 12

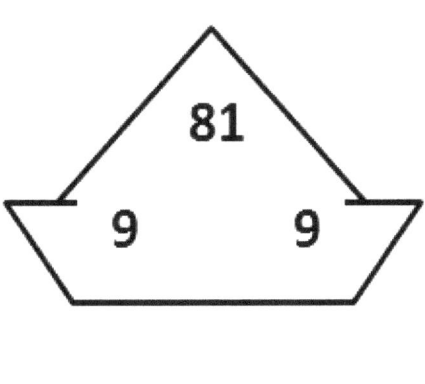

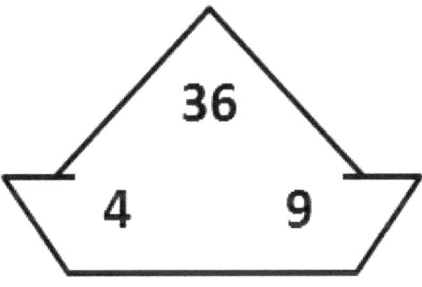

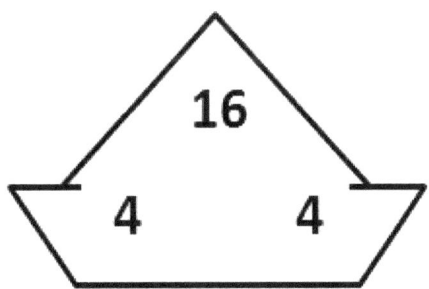

END NOTES

[1] The 'system' is not a monster; it can be thought of as an evolving and self-organising set of rules. It is like an adapting organism. Sometimes it doesn't adapt well and one or more generations are left to deal with the consequences.

[2] I have spoken to many people from an older generation who speak fondly of those latter-day school masters who used the cane to 'encourage' their students to learn. Their former pupils made distinctions between those who did it through a sense of duty to the child and those who resorted to physical punishment as a convenience. They appreciated the former and resented the latter. We judge actions by their intension; so do children!

[3] If you search the internet for 'dumbing down of education' you will find that there is a great deal of debate going on about this issue and, in most cases, it is *not* about whether it is true that state education has dumbed down over the last few decades, but more about the possible reasons for this. Here, I am debating that there is another factor that needs to be taken seriously; that children should learn (ok, memorise) some things whether or not they understand them... *yet*. However in many education circles such a notion is so incomprehensible that it is seen as being akin to violating the children's human rights. I beg to differ on this point.

[4] I was tempted to start this paragraph with, "As children, *we* have an enormous potential..." to highlight the fact that the child that we once were never leaves us entirely. But since this book's primary focus is on what makes adults and children different, I decided to curb my enthusiasm to point out that we should, wherever possible and appropriate (i.e.

not always, but not never either), treat children as we would have liked ourselves to have been treated, that is, the treatment that would have led to the *outcome* that we would have wanted for ourselves, not the fleeting feeling that we would have wanted to have felt at the time.

[5] And, at worst, fantasy and fictional films and stories that fill our children's heads with non-existent and, at least non empowering, creations of marketing professionals' imaginations. Note that, legends, on the other hand, are in quite a different category. However, I am not going to open up that Cyclopean topic in a book like this.

[6] Yes, of course, you can say that any language is about relationships. For example, the word 'island' describes the relationship between 'land', 'sea' and the notion of 'surround' in such a way that the 'sea surrounds the land' and the reverse relationship would be called a 'lake'. But language is not designed to specifically answer questions beginning with, "What is the relationship between …" and mathematics is. Also, the language of mathematics is universal.

[7] Many philosophers may disagree with me on that point, but I am confident that any philosopher who is also a competent mathematician is very likely to think, write and propose new ideas that are qualitatively different to his or her innumerate colleagues. The use of the word innumerate here was inspired by a book called 'innumeracy' by John Allen Paulos.

[8] Over a thousand years ago, a father, who died when the boy was still in his early twenties, provided his son with the best education that money could buy, which at the time meant seeking out those who were renowned for their prowess and getting them to teach his son privately from the age of five. In spite of being persecuted for his father's religion and moving from city to city whilst war was being

waged all around him, his renown as a scholar meant that he found patrons everywhere he went. Who would have thought that this young man would transform Islamic philosophy and yet, be best known as 'the father of modern medicine' to whom the world is indebted for its current level of heath and consequent prosperity? I suggest that you go read about him with the intention of being inspired. Every one of our children has the potential to be an Avicenna, if only we provide them with the will, and the candlelight (not LCD or Plasma light :-). And do you think you could imagine what the world might be like now, if every one of us and our children had been given that opportunity.

[9] I first came across what later became known as the Butterfly Effect when, back in 1976, my English teacher read 'The Sound of Thunder' by Ray Bradbury for us. Other depictions of this effect can be found in the URL at the end of this sentence, the lessons for us being that, whatever we do, or avoid doing, however trivial it might seem, has repercussions beyond our comprehension.
http://en.wikipedia.org/wiki/Butterfly_effect_in_popular_culture

[10] You might argue with me here and say that the system has worked for generations, so why should we rock the boat? And my answer would be that if we thought like that we would still be riding on log rafts and pushing them along with long sticks. There's nothing intrinsically wrong with log rafts, of course. In fact, a few thousand years ago, the way log rafts were constructed could make or break empires. But we need to move with the times and … the times tables.

[11] By which I mean, "most effective", by which I mean, "most empowering", by which I mean, "enabling your child to feel competent and confident", by which I mean that "your child will enjoy doing them and be proud of his

achievement and will be able to use what she has learnt to improve her quality of life."

[12] I once had a student whose mother told me that he used to be very good at maths, but had stopped trying. When I asked him why, he said that he wanted to be a Philosopher and, as such, what use would mathematics be to him? After I explained that, just like philosophy, mathematics is also concerned with relationships, only using a different language to access those aspects of relationships between phenomena that everyday language is not so good at, he became motivated to do well at it.

[13] Take out your calculator if you like, but I quickly said to myself, that's 320+40, which is 360. But it doesn't matter. The calculator is there, so use it if you prefer. It may be too late for you anyway, but it's not yet for your children ☺

[14] Looking at the bigger picture we could see the problem from the perspective of the contrived transition between what is perceived to be primary and secondary school. This change of focus also happens at the next 'level' where mathematics and physics professors and lecturers at universities disagree with students' perceptions of the nature of the subjects as they enter university. Many students drop out of university because their perception of what the subject was going to be like is different to what they actually encounter.

This problem can be traced back to 'primary' mathematics often being taught by teachers who are not expected to have any real appreciation of higher level mathematics. A primary school teacher who considers mathematics to be the manipulation of number, would not be concerned with the process by which the 'answer' is arrived at; only that, eventually, the child is able to apply the 'principles' to solve numerical questions. This gives children a concrete, arithmetic, view of mathematics, whereas higher level mathematics *requires* an abstract, that is, a 'variable' or 'algebraic' perspective. A primary school teacher who knows that the arithmetic is not an end, but a tool that is going to be at the service of algebra [a rare breed of primary school teacher], would expect multiplication facts to be memorised and not 'worked out' so that children would be able to focus on seeing patterns that can be generalised and not whether or not the child can, eventually, arrive at a solution to a simple arithmetic problem.

The issue is that by calling both levels (primary and secondary) 'mathematics', we are implying that higher level mathematics is an extension of primary school mathematics. This can create expectations in students that does not prepare them for the different way of thinking that is required at higher levels. This also happens when students move from secondary school to study mathematics or a related subject at university. The high attrition rate in students starting degree level mathematics attests to the chasm between what is expected and what is experienced by students upon entering university to study mathematics.

I have noticed a similar attrition in interest and motivation when children progress from primary school to secondary school (although, for reasons I am about explain, I have no numbers to back this up – If this hadn't been an end note, I would made this bit a footnote!). However, this is not so obvious because the children cannot 'opt out' of mathematics pre-16 and so the disillusionment is not picked up by the existing tracking processes.

[15] Just like any other language, the language of mathematics serves to answer one of three questions; "What is it?", "What does it mean?" and "What can I use it for?" . That is, a) to *describe* (i.e. communicate) a relationship and b)

to *derive* new relationships and c) assign meanings (i.e. understandings) to those relationships, based on what is already known. In other words, we first use language to understand existing relationships and *then* to create new ones from those. In mathematics, the first purpose often leads to what we call 'describing patterns' and the second purpose is called mathematical 'problem solving'. AND you can't do any of these, unless you know the language (you know, the basic arbitrary stuff that doesn't make any sense, like why we write an A like this [a] or this [α].

[16] If you are interested (and I urge you to be at least curious about this), this idea of psychological 'defence mechanisms' was proposed by the famous Sigmund Freud himself. Here's one possible source of information on this topic: https://en.wikipedia.org/wiki/Defence_mechanisms

[17] I urge you to find out more about 'confirmation bias' too. Just Google it - I know that's not politically correct and I should say, "use a search engine to find it," but then nobody complains if someone calls a vacuum cleaner a 'hoover' (yes, with a small 'h'), do they?

[18] There's something else that I cannot go into detail here, which is that 'confidence' (or to be more precise, 'self-efficacy', if you want to delve deeper into the psychology of it) is very closely linked to our perception of our social position. Top students in bottom sets feel more confident than students at the bottom of top sets. In education circles, thanks to Herbert Marsh, we call this the "big fish little pond effect." I have seen this happen. Students who were performing well above average in a non-selective school move to a selective school where their performance is average and they begin to lose their self-confidence. I have also seen the reverse, where an 'average' student with low self-confidence moves from a school with a strong emphasis on academic performance to

a school with emphasis on effort and encouragement and becomes more confident and more interested in her own academic performance and achieves more than before, in spite of being in a less academically challenging environment. So, here's the caveat: Be careful - our children's 'grades' depend on a great many factors; not just their so-called 'ability', or even effort. I have put the word 'ability' in quotation marks because we cannot measure ability directly, we can only devise (often very crude and inadequate) methods to help us *infer* ability and we call the results of those measures 'attainment'.

[19] Especially since they can't be explained because they are arbitrary anyway, although that doesn't stop my daughter from asking questions like, "Why is the letter 'a' written the way it is?"

[20] You would have to forget about the qualitative difference between the marbles to do the maths, which means that you would have to forget that they are marbles and just concentrate on the numbers, which is the same as losing sight of what makes the marbles, marbles. So, I am not far off when I say that the rest of us are 'losing our marbles' - and everything else that we reduce to mere numbers, albeit for convenience. This reminds me of a section in 'The Little Prince' by Antoine de Saint-Exupéry:

> *"Grown-ups love figures... When you tell them you've made a new friend they never ask you any questions about essential matters. They never say to you "What does his voice sound like? What games does he love best? Does he collect butterflies?" Instead they demand "How old is he? How much does he weigh? How much money does his father make?" Only from these figures do they think they have learned anything about him."*

[21] We are expected to discover this for ourselves, and we generally do. Just as we are expected to understand the

concept of zero straight away (and surprisingly enough, we do), when in fact the first mention of it in history is found in an encyclopaedia written by the Persian mathematician, Khwarizmi, as recently as just over a thousand years ago (976 CE to be precise). In other words, it took Man thousands of years to come up with the idea of zero that we now so readily accept when it is presented to us.

22 We are so used to thinking in abstract or metaphorical ways, that we often find it amusing or strange when what we say is taken literally. And sometimes it seems so obvious to us that we are baffled by a response that we get that is based on literal interpretation, having completely forgotten that there could be a literal interpretation which would then make the response quite logical. For example, I had not been in the UK for very long when, at school, one of my teachers said to me, "Bijan, I am only pulling your leg." I looked at him, then I looked at my leg and I could not understand how he could have done this without me noticing. Notice that the significance of the word 'only' here escaped me (I ignored it) just as the children playing with the marbles ignored the word 'number' when they were asked whether or not the number of marbles in each box was the same.

23 As a rule, the more we *identify* with something, the more motivated we are to know more about it. And we tend to identify very much with what we *possess*: *my* car, *my* book, *my* child, *my* knowledge. So, when your child knows *her* times tables, it is *hers*. As such, she is more likely to identify with it and want to more about it and how to use it. Another way of looking at this is to say that our minds become more sensitive to noticing (filtering in) what it identifies with, making it more likely for someone who knows their times tables to notice opportunities for using it.

[24] 'Concrete' refers to a kind of thinking process which says that, if something is not tangible, it cannot be understood. In other words, if you can't see it, touch it, hear it, smell it or taste it, such as the number five without five things to represent it, there's nothing there to think about. The child will not be able to create a representation of the number five as an 'abstract' idea, but only as five objects. Similarly, for children in the early concrete stage (very young children), if someone goes behind a wall, that person, *literally*, disappears.

[25] I am using the word memorise instead of 'learn' deliberately because, for some of my educationalist colleagues, if something that is experienced does not lead to greater understanding, it cannot be considered to be learning and I am trying to avoid that possible complication.

[26] Although it could be argued that there is really no need to invoke the concept of trust in here since the child is simply doing what he's asked as a means to an end; getting attention and praise.

[27] Actually, to young people nothing is pointless. If you give them Pokemon cards, they will fill their amazing brains with the name of every character and its so-called abilities irrespective of whether or not this has any real value. So, why not give them a book about how to recognize different plants by their characteristics and perhaps their medicinal properties or animals and their habitats or rocks and their geological significance or the elements and their features? That way, they will fill their brains with information that will help them appreciate the wonders of nature when, *later*, they begin to try to *understand* the relevance of their knowledge to themselves, their community and their future. The periodic table of elements would also give them an advantage, since familiarity with the symbols and

the structure would help them to engage better with chemistry when the time comes.

[28] Identifying a missing smaller number is, in effect, an exercise in division, even though the child will not be able to recognise it as such, yet. For example, having memorised the 6,8,48 triad, if the child is presented with the numbers 8 and 48 and can identify that 6 is the missing number, then, when the time comes (probably much later) all the child needs to do is to be told that the process of obtaining the larger number from the two smaller ones is called multiplication and if one of the smaller numbers is missing, then the process is called division.

[29] I should make a distinction here between speed and reliability of recall. Associating meaning to items to be recalled *does* improve the *reliability* of what is recalled, however, it is a slower process. Therefore, to increase reliability, we need to compensate for the lack of meaning anchors and we do this through repetition or meaningless anchors, such as rhythm and music or visual associations.

[30] OK, they might eventually. But in most cases they give up before they become proficient because it is not an efficient way of doing things. Since the child has to look up every single fact that he needs to solve, even simple multiplication problems, students are put off mathematics because taking so long to obtain the answer makes the child tired and frustrated. And since at this stage this is all that they perceive mathematics to be, they begin to become disillusioned with 'mathematics'. How much would a child lose out if they became disillusioned with reading and writing because they found looking up every word that they need to spell boring or frustrating? I can understand why. They want to express themselves and by the time they have found out how to spell the word, the idea will have escaped them, as it would do with most of us (I see

this with my own children. They were, at one point, reluctant to write because they felt that it showed up their weakness in spelling rather than highlight their strengths in being able to express their ideas).

31 Regrettably, more commonly, the intention is to make the ideas easier to test, often more for the benefit of administrators than the children themselves. This is the basis for the big debate in educational circles between 'formative assessment' (for the benefit of students and therefore, involves giving feedback to students with a view to helping them improve) and 'summative assessment' (for the benefit of the 'system' to compare and categorise children, teachers and schools).

32 Because of the way in which I learnt my times tables, whenever I am faced with a problem requiring simple division, I have to do a reverse search through my times tables. For example, when faced with 24/6, I (albeit fairly quickly nowadays) have to subvocalize "six, four, twenty four" to confirm that I have the right answer. This method gets around this problem.

Have fun with your kids and remember to give them plenty of **attention and praise.**

For news, updates, worksheets, discussions and further mathematics resources, please visit:

www.mathtery.com/tt

I would love to hear about your experiences with this method. So, feel free to connect with me through the website.

> *Please note that there is **patent pending** on the "**Triad method**" described in this book and you're required to obtain a license if you wish to make commerial use of (make money from) the concept. Feel free to contact me about this via the above website if you are interested in doing this.*

The Blurb

This book is essential reading for everyone concerned with mathematics education, including parents, grandparents, teachers, tutors, university lecturers and policy-makers. It explains one of the fundamental reasons why children fail to fall in love with mathematics (or not), leading to lost opportunities, not only in terms of wasted potential that could benefit society as well as their own academic and career achievement, but also in relation to their, and consequently everyone's, greater life fulfilment.

The times tables has a crucial role in our children's journey towards that more inspired world view that mathematical understanding bestows. As parents, teachers and/or policy-makers, we need to appreciate how important it is for children to memorise their times tables ***efficiently***. Only when we, as guardians of all our children's futures, make the efficient memorisation of the times tables a priority, will questions about *how* our children memorise their times tables become a matter for scrutiny, and *when* they memorise their times tables become a matter of urgency.

In this book, Dr. Riazi-Farzad also explains why calculators do not replace the need for us to have rapid mental access to the times tables.

Making a strong case for why most of the current methods for teaching the times-tables are highly inefficient, Dr Riazi-Farzad offers an alternative (patent pending) method of introducing the times tables to children that will help them to, not only recall their times tables facts efficiently, but also provides a seamless transition to division and beyond.

Dr. Bijan Riazi-Farzad has been teaching science and mathematics since 1990. After studying asthma for his PhD in Pharmacology, he turned his attention to education and obtained a Masters degree in Psychology of Education in 2006. Since then, he has been involved in research at the Institute of Education, University of London, looking into factors that affect young people's perception of their school subjects, particularly Mathematics and Physics. He has two primary-aged children and lives in London, United Kingdom.

www.ingramcontent.com/pod-product-compliance
Lightning Source LLC
Chambersburg PA
CBHW051546170526
45165CB00002B/903